Dr. Strangelove's Game

By the Same Author

Science
Pythagoras and His Theorem
Archimedes and the Fulcrum
Galileo and the Solar System
Newton and Gravity
Darwin and Evolution
Curie and Radioactivity
Einstein and Relativity
Bohr and Quantum Theory
Turing and the Computer
Oppenheimer and the Bomb
Crick, Watson and DNA
Hawking and Black Holes
Mendeleyev's Dream

Philosophy
Philosophers in 90 Minutes Series (twenty-four titles)

Novels
Pass by the Sea
A Season in Abyssinia
One Man's War
Vaslav
The Adventures of Spiro

Dr. Strangelove's Game

A Brief History of Economic Genius

Paul Strathern

ALFRED A. KNOPF CANADA

To Tristan and Julian

PUBLISHED BY ALFRED A. KNOPF CANADA

Copyright © 2001 Paul Strathern

All rights reserved under International and Pan-American Copyright
Conventions. Published in 2001 by Alfred A. Knopf Canada, a division
of Random House of Canada Limited, Toronto. Distributed by
Random House of Canada Limited, Toronto.

Knopf Canada and colophon are trademarks.

National Library of Canada Cataloguing in Publication Data

Strathern, Paul
    Dr. Strangelove's game : a brief history of economic genius

ISBN 0-676-97448-1

1. Economists — Biography. 2. Economics — History. 3. Economics —
Philosophy. I. Title.

HB76.S77 2001   330'.092'2   C2001-930749-7

First Edition

www.randomhouse.ca

Set in 10/13 pt Caslon 540
Typeset by Rowland Phototypesetting Ltd, Bury St Edmunds, Suffolk

Printed and bound in the United States of America

10 9 8 7 6 5 4 3 2 1

Contents

List of Illustrations

Every effort has been made to contact copyright
holders. The publishers will be glad to make good in
future editions any errors or omissions brought to their
attention

List of Figures

'The ideas of economists and political philosophers, both when they are right and when they are wrong, are more powerful than is commonly understood. Indeed, the world is ruled by little else. Practical men, who believe themselves to be quite exempt from any intellectual influences, are usually the slaves of some defunct economist. Madmen in authority, who hear voices in the air, are distilling their frenzy from some academic scribbler of a few years back.'

John Maynard Keynes

'In economics there are some, even if not many, immutable laws – laws of an order of certainty of Calvin Coolidge's possibly apocryphal dictum that when many people are out of work, unemployment results.'

J. K. Galbraith

Prologue

Dr Strangelove had a black-gloved synthetic arm; his crippled body was confined to a wheelchair. Shaded glasses obscured his intense twisted features, and his high-pitched, strangulated voice had a menacing mid-European accent. This was the evil genius of the War Room, advising the US president on strategy as the world faced nuclear disaster, in Stanley Kubrick's film *Dr Strangelove, or How I Learned to Love the Bomb*.

A sole American bomber has eluded the Soviet defence system, and is now beyond recall. With horror, the Soviet ambassador reveals to the president and his advisor the likely consequences: if the bomber succeeds in reaching its target, it is liable to trigger the Soviet Union's ultimate weapon, the Doomsday Machine. This will release a vast cloud of radioactive material which will enshroud the entire Earth, destroying all human and animal life for 100 years. The president is aghast. Dr Strangelove is exasperated, and exclaims to the Soviet ambassador, 'The whole point of hafing a Doomsday Machine is lost *if you keep it a secret. Vy didn't you tell the vurld?*'

This is the logic of game theory – the first explicit reference to this new method of strategic thought in a popular movie. The entire notion of nuclear deterrence was based upon game theory.

Dr Strangelove, somewhat frenetically overacted by Peter Sellers, shows signs of growing insanity as the film approaches its nuclear climax. He rapidly conceives of a brilliant plan for human survival, his mechanical arm starts going out of control and attempts to strangle him, he makes agitated references to a 'kom-pew-tah'. The evil genius degenerates into a parody of the mad scientist.

Most people considered Dr Strangelove to be a far-fetched creation, exaggerated for satirical purposes. The facts suggest otherwise. During the mid 1950s a mysterious figure with a Hungarian accent, his crippled body confined to a wheelchair, would be whisked by limousine from his bed at Walter Reed Hospital in Washington to the White House. Here, President Eisenhower, who had previously commanded the entire Allied forces in Europe during the Second World War, would listen intently to the suggestions of his secret strategic advisor, a man who had never even been in the army. The meeting over, the wheelchair-bound figure would be sped back to his hospital room. Here, two armed guards were posted at his door night and day, and he was attended only by naval nurses with top-security clearance. The patient was becoming increasingly deranged and would frequently wake shrieking and babbling in the night. His military minders were there to ensure that any secrets he blurted out would not find their way to a foreign power.

This was John von Neumann, a brilliant Hungarian who was responsible for major breakthroughs in several intellectual fields, from pure mathematics to practical economics. Indeed, in 1944 he was convinced that he had found a method which 'solved' economics. From then on 'wise' economic choice would simply be a matter of mathematical calculation. The entire process of economic decision-making could be left to computers (which he also helped to invent), and economists of the human variety would thus become redundant. This same man – arguably the finest mathematical mind of the twentieth century – also proposed even more far-reaching uses for his radical new method, which he called game theory. Here was a theory which showed not only how to banish for ever economic uncertainty, but also how to rule the world by nuclear force.

John von Neumann was born the son of a rich banker in Budapest in 1903, one of the brilliant Hungarian generation which was to produce figures ranging from Georg Solti to Zsa Zsa Gabor. Von Neumann was quickly recognized as an infant prodigy. By the age of six he could read a page from the Budapest telephone directory once and immediately repeat it from memory. By the time he was eight he could divide two eight-digit numbers in his head. (Try dividing 97,572,915 by 18,835,769 *on paper*.) Before the age of thirty, von Neumann wrote what came to be regarded as the definitive textbook on quantum mechanics. But this work contained von Neumann's first crucial mistake: an erroneous proof. Yet by now his reputation was so great that the few who spotted this error felt sure they must have overlooked something. The force of von Neumann's logic was incontestable, but it had been based upon an unjustified assumption. As we shall see, this was to become a characteristic flaw. (Von Neumann's 'proof' would hold back a particular aspect of quantum mechanics for over half a century.)

In 1928 von Neumann came up with a theory which was to transform the long history of mathematical probability. This would become known as game theory. Its intention was to reduce any two-person contest to a precise mathematical game. A player's alternatives in a poker game could be assessed according to the mathematical probability of their different outcomes. However, game theory was about more than games; it could also be applied to reality. In von Neumann's words, 'Real life consists of bluffing, of little tactics of deception, of asking yourself what is the other man going to think I mean to do. And that is what games are about in my theory.' Game theory was about conflict between two highly intelligent and deceitful partners, conjoined within certain rules. One player could never be sure whether the other was double-crossing.

Von Neumann's continuing obsession with game theory may well have been linked to his compulsive sex drive. According to his biographer, Steve J. Heims, 'some of his colleagues found it disconcerting that upon entering an office where a pretty secretary was working, von Neumann habitually would bend way over, more or less trying to look up her dress'. Von Neumann continued womanizing throughout his two marriages. Both of his wives were powerful and intelligent women, who proved quite capable of double-guessing what even a genius might get up to in his spare time. In a letter to his second wife after a particular misdemeanour had been discovered, von Neumann wrote guardedly, 'I hope you have forgiven my modest venture in double-crossing.'

In the light of such behaviour, it is not surprising to find that the inventor of game theory took a somewhat cautious view of two-person games and conflicts. There was only one rational strategy: 'defeat is inevitable if you aim to win rather than avoid losing'. Damage limitation was the object of the exercise. At every stage, you should work out each possible move that you can make, and then calculate the maximum possible loss that you could sustain if you made that move. You should then select the move which had the minimum maximum possible loss. This became known as minimax theorem – and though it resulted in one divorce, it seems to have minimized the possibility of a second maximum loss for its creator.

Von Neumann emigrated to the US in the 1930s. Not yet thirty, he was appointed alongside Einstein to the newly founded Institute for Advance Study at Princeton. The Institute, which was devoted exclusively to theoretical research, quickly became a Mecca for the finest scientific brains in Europe and America. While at the Institute, von Neumann would later play a leading role in the development of the first computers. Despite the ban on all practical work at the Institute, he managed to assemble in the boiler room a large prototype computer, an early version of the Mathematical Analyzer, Numerator, Integrator and Calculator (known as MANIAC).

Von Neumann also made vital contributions to the development of the first nuclear bombs. As a result, he became the leading member of the US Atomic Energy Commission, advising the president on the use of the hydrogen bomb. The Cold War against Soviet Russia was now entering its most frigid period. Von Neumann saw this global conflict as an ideal opportunity for putting game theory into practice. Here was the two-person game to beat all two-person games. The application of game theory to this situation led von Neumann to but one conclusion, the logic of which was incontestable. The only possible course of action was to strike first. As von Neumann earnestly explained to the president, game theory dictated that he drop an H bomb on the Russians at once. The only way to maintain a winning position in the game was to destroy the Russians before they could develop their own H bomb. Any other course of action would be completely contrary to the logic of game theory. (Never before, even at home, had he come across such an opportune game situation.)

Once again, we can see that von Neumann's logic is irrefutable. However, his assumptions appear to have been subjected to somewhat less rigid analysis. Regardless of von Neumann's urgent insistence ('bevor it is too late'), President Eisenhower continued to hesitate. Even his secretary of state, John Foster Dulles, became convinced by von Neumann's logic. But still Eisenhower dithered. He found himself unable to counter the compelling argument put forward by these two wizards of global strategy. Despite this, the man who had won the Second World War in Europe couldn't help feeling that something was amiss. Then the Russians announced that they also now had the H bomb. It was too late. Despite the illogic of its position, sanity had prevailed.

But game theory could be applied to situations that were more than just catastrophic (nuclear strategy, divorce) or trivial (poker, marriage). By its very nature, it referred to any human activity involving conflict. This meant that game theory could be applied to the most complex and vital human activity of them all, namely, economics. As early as 1939 von Neumann had been approached by another brilliant Austro-Hungarian member of the Zsa Zsa Gabor generation with just this view in mind.

Oskar Morgenstern had studied economics and philosophy in Vienna. This had led him to conclude, 'I was an idiot [to have studied] this silly philosophy.' His opinion of economics, or more precisely economists, was little better. He dismissed the work of the leading Austrian economist Friedrich von Hayek as 'higher nonsense'. Despite this, he was willing to succeed Hayek as director of the celebrated Vienna Institute for Business Cycle Research. Morgenstern's ability to antagonize all and sundry came to a head with the Nazi takeover of Austria in 1938. Though anti-Semitic, Morgenstern was viewed as 'politically unbearable' by the Nazis. He accepted a post on the economics faculty at Princeton, and remained in exile for the rest of his life. His American colleagues proved an equal disappointment: 'Economists simply don't know what science means. I am quite disgusted with all this rubbish.'

Not content with his low opinion of others, Morgenstern compounded this with delusions of his own grandeur. He claimed to be an illegitimate descendant of the German Kaiser Friedrich III, and kept a portrait of his 'grandfather' on prominent display in his apartment (even after America went to war with Germany). Morgenstern soon became a recognizable figure about Princeton. Wearing one of his tailor-made three-piece outfits, he took to riding on horseback through the streets.

In the words of the American economic historian Robert J. Leonard, Morgenstern was a man of 'enormous intellectual ambition and limited theoretical ability'. He wished to mix only with those he considered to be his intellectual peers, and quickly gravitated to the Institute for Advanced Study. Here, the likes of Einstein, Gödel and the resident Nobel prize-winners soon learnt to avoid him. Not so von Neumann, who established a curious rapport with Morgenstern. It seems Morgenstern was willing to overlook von Neumann's Jewishness – possibly on account of his titled 'von', as much as his intellectual pre-eminence. Morgenstern's view of economics was shared by von Neumann, who declared, 'economics is simply a million miles away from the state in which an advanced science is, such as physics.'

Morgenstern saw his chance. Here was just the man to make up for those minor theoretical deficiencies which so often seemed to hamper the full flowering of his exceptional intellect. If only he and von Neumann could collaborate on a project. Morgenstern began looking through von Neumann's previous work and came across his paper on game theory. Yes, this was it! Here at last was the inspiration they had both been waiting for, he assured von Neumann. With his economic genius, and von Neumann's mathematical acumen, they would together rescue economics from its neolithic state. They would transform it into an exact science, whose incontestable logic would leave no possible room for error. This would be done by the application of game theory.

In 1941 Oskar Morgenstern and his new pal 'Johnny' von Neumann began collaborating on a paper which demonstrated how game theory could be applied to economics. This paper soon grew into two papers, and then blossomed into a 100-page pamphlet. Despite war shortages, Princeton University Press was persuaded to publish – as soon as the authors completed their final draft. But by now Morgenstern and von Neumann had become obsessed with their topic, and the pamphlet soon began to expand into a book. At this stage von Neumann was doing stints at Los Alamos, where he was working on the Manhattan Project to build the first atomic bomb. He was also in demand as a consultant to the military and the government in Washington. But in between the distractions of advising how to win the war and devising how to split the atom, von Neumann would return to Princeton to continue with his collaboration. He and Morgenstern worked night and day. Von Neumann would rise while his wife was still asleep and meet Morgenstern at the Nassau Club. Here, they would review their previous work over breakfast and then sketch out further possibilities for the expansion of game theory into economic activity. Discussion would continue at Morgenstern's apartment across Nassau Street above the Princeton Bank. Under the watchful eye of the Kaiser, Morgenstern would scribble intently in his pad as von Neumann gazed into the middle distance, spouting a succession of abstruse mathematical formulae. (The child who was capable of dividing eight-figure numbers in his head could now conjure up formulae which reduced all economic activity to a game.) In the afternoon they would retire to von Neumann's house at 26 Westcott Road, so that he could spend some time with his wife Klari during his brief spell on leave from Los Alamos and Washington. Klari would serve coffee as the two collaborators continued with their conversation. The talk would continue over more coffee, over drinks, over dinner, and yet more coffee. Having been ignored for eight hours at a stretch, Klari would finally turn them out of the house, where they would walk the night-deserted streets of wartime Princeton until the early hours. Eventually Morgenstern noticed, 'Klari was often rather distressed by our perpetual collaboration and incessant conversations.' So at the weekends they would spare her this upsetting experience. Instead, 'we drove occasionally to the seashore and walked up and down the boardwalk at Sea Girt, in particular, discussing matters'. In the end they all travelled over a thousand miles to the Texas coast on the Gulf of Mexico: 'I was vacationing in Biloxi with Johnny and his wife Klari. Again, day after day was spent in discussing the theory ... Incidentally, we always spoke German.' One can only wonder what the locals made of all this. At the time the entire eastern seaboard was rife with rumours of Germans recording the movement of Atlantic convoys to Europe, and stories about enemy spies being put ashore from submarines.

By April 1943 their work was at last complete. As Morgenstern explained, 'The people at the Press were quite overwhelmed seeing a manuscript of about 1,200 typed pages full of graphs and uninhibited mathematical notations.' Without apparent irony, its authors decided to call this magnum opus *General Theory of Rational Behavior*. But eventually they felt this 'was not descriptive enough of our work', and it was retitled *Theory of Games and Economic Behavior*. The work is prefaced by an assurance that 'no specific knowledge of any particular body of advanced mathematics is required. However . . .' A glance at the ensuing hundreds of pages of tightly packed formulae containing many 'uninhibited mathematical notations' should suffice to indicate whether your mathematics rises to this humble category. The work itself opens with a modest comparison between the application of game theory to economics and the effect of Newton's discovery of gravity on physics. It passes on to analyse such diverse subjects as the 'Robinson Crusoe' economy, 'The Adventures of Sherlock Holmes' and 'Poker and Bluffing', coming to a close with 'Economic Interpretation of the Results' for markets consisting of two people, and extending even as far as markets 'greater than three people'. At the end there is an index which includes such items as 'pyschological phenomena, mathematical treatment'; 'winning' has sufficient entries to include two sub-sections: 'certainly' and 'fully'. There are just two entries under 'losing'.

The publication of *Theory of Games and Economic Behavior* in 1944 was greeted with ecstatic acclaim. The *American Mathematical Society Bulletin* described it as 'one of the major scientific contributions of the first half of the twentieth century' – thus placing it alongside relativity, quantum theory and the Keynesian economics which had combatted world recession (though Morgenstern was of the opinion that 'Keynes is a scientific charlatan, and his followers not even that'). More cautious critics like the economist Leonid Hurwicz still remained hugely optimistic: 'Ten more such books and the future of economics is assured.' The groundswell of opinion in favour of *Theory of Games and Economic Behavior* rose to a climax in March 1946, when the book made front-page news in *The New York Times* – on a par with Churchill's declaration in the same month that an 'Iron Curtain' had descended across Europe. Many were convinced that game theory would become the foundation of all economic theory. Economics would be reduced once and for all to mere calculation. There would be no room for human argument: a decision would be either right or wrong.

Dr Strangelove's game could have put an end to the world as we know it. Might it one day do something similar to economics? The answer lies in our definition of economics. Initially the subject consisted of little more than a series of insights concerning commercial activity – what exactly it was, how it worked, how to improve upon it. From such beginnings the study of economics gradually evolved. Over the centuries, this has generated an epic narrative, replete with a cast of varied and colourful personalities. But instead of a traditional story, with a central hero whose actions illustrate and develop his character, we have a central idea. This evolves in the hands of a succession of individuals, including some of the greatest thinkers of their age, as well as a motley collection of wayward geniuses, moralists, eccentrics and charlatans. Some have attempted to save the world, and might have destroyed it. Some have seen themselves simply as mechanics, doing their best to keep the engine running. Others have conjured up utopias or apocalyptic visions. A number have been drawn by compassion to try to remedy the horrific reality they saw unfolding before them. The ever-evolving economic idea has not always resulted in progress – far from it. Yet it has led to a deepening social self-understanding, and a growing insight into how society works. From an initial series of insights, economics has now spread into every facet of our lives. It has made us increasingly aware of who we are and what it is we are doing. What follows is a narrative of how this happened, together with the lives and ideas of those who helped to make it happen.

I

Something Out Of Nothing Comes

That telling cultural symbol, the zero, first arrived in Europe from the Levant around 1200. Prior to this, calculation had best been carried out on the abacus. With no concept of zero, it had been mathematically impossible to conceive of negative numbers. With the advent of the zero, more complex commercial calculations could be carried out on the page, and you could end up with a minus quantity. Accountancy as we know it had been born.

By the late thirteenth century merchant banks were conducting international trade throughout Europe and beyond. Ledgers in Brugge recorded batches of sealskins from Greenland to pay papal dues. Marco Polo (or the genuine sources he used to confect his memoirs) observed Genoese ships trading on the Caspian Sea. The first great European bankers were the Bardi family of Florence, who flourished from 1250. By the following century they had risen to became the sole papal bankers, holding a monopoly on the collection of papal revenues throughout the Continent. Such was their wealth, they could even afford to finance kings and wars. It was their support for the extravagances of Edward III, who led England into the Hundred Years War with France, which eventually precipitated their downfall. When Edward reneged on his colossal debts, there was nothing the Bardis could do about it. Commerce had money and influence, but as yet no real power. In 1345 the House of Bardi went bankrupt, contributing to an economic downturn throughout Europe.

Two years later a trading ship from the Black Sea drifted into the Bay of Naples, its sails slack, its ropes unmanned. Aboard, the crew were found to be either dead or dying in agony, their bodies disfigured with black pustules and grotesque swellings. The Black Death had arrived in Europe. Within four years 25 million of the Continent's inhabitants would die – more than a third of the entire population.

It would be over a century before European trade fully recovered. However, by then commerce had been transformed by an invention whose importance was likened by contemporary thinkers to the discovery of mathematics or the recent invention of the printing press. Like the zero, double-entry book-keeping seems to have arrived in Europe at Genoa from the Middle East. Again like the zero, it was probably the product of Hindu-Arabic mathematics, adapted to the hectic and far-flung trading of the bazaars of Baghdad.

The man largely responsible for the spread of double-entry book-keeping through Europe was Luca Pacioli, the forgotten genius of the Renaissance. This was the Italian monk whom even Leonardo called 'maestro'. Pacioli was born around 1445 in the mountains forty miles east of Florence at Sansepulcro. Significantly, this was also the home town of the artist Piero della Francesca, whose understanding of perspective transformed Renaissance painting. Piero viewed perspective as a mathematical science. His paintings are an explication of geometric shape and line: mathematical space imbued with the particularity and plasticity of realistic scenes. Indeed, such was his enthusiasm for mathematics that for the last decades of his life he abandoned painting altogether. He returned to his native Sansepulcro to pursue his obsession. It was during this period that Pacioli studied with him, his able young mind fired by Piero's enthusiasm.

At twenty Pacioli left for Venice, where he entered the service of a prosperous merchant called Rompiasi. Pacioli tutored Rompiasi's sons and 'on account of this merchant I travelled in ships carrying goods'. How far he travelled is not certain, but at this point the

*Luca Pacioli, the monk who popularized
double-entry book-keeping and taught Leonardo
maths*

Venetian empire stretched as far as Cyprus, and its merchant
galleys traded regularly with Beirut, the terminus of the silk cara-
vans from China and the spice route from India. So it's quite
possible that the young Pacioli travelled as far as the Levant. In
the manner of the great mathematical thinkers of ancient Greece
(such as Pythagoras and Plato), Pacioli too may have acquired
some of his mathematical expertise directly from Eastern sources.
Though with the Renaissance burgeoning in Italy, European
mathematics was now beginning to outstrip its Arabic benefactor.

More importantly, Pacioli not only studied mathematics but was also involved in trade.

Around the age of thirty, Pacioli became ordained as a priest in the Franciscan order. There is no doubting his firm belief in God, but his reason for taking monastic vows remains obscure. His relationship with his superiors was always uneasy. Financial security and access to teaching posts at universities would appear to have been at least partly his motive for committing to the monastic life.

From now on Pacioli lived a restless existence, teaching for varying periods at universities all over Italy – including the best, at Padua and Naples. He also published several mathematical treatises during this period. These contain little original work, but exhibit in succinct and profound form an unmatched, encyclopedic knowledge of his subject. However, one originality may have been the so-called 'problem of the points' which he describes.

'Two players are engaged in a fair game of balla,' Pacioli explains. Balla was a medieval ball game, popular as a form of gambling; the insertion of the word 'fair' would seem to have been necessary given the place and period. 'The two players agree to play on until one has won six rounds. The game is interrupted when one player has won five rounds, and the other three. How are the initial stakes to be divided?' This seemingly innocent puzzle in fact posed a problem of extreme subtlety, open to differing solutions. Not until a century and a half later would the two great French mathematicians Pascal and Fermat (of Last Theorem fame) see the full mathematical implications of this problem which had several answers. Here it was possible to grasp, in numerical terms, the future. The player who was ahead in the game of balla was more likely to win, but his victory was far from certain. Mathematics could deal with all possible future outcomes, and calculate the odds on each of these taking place. The calculation of these odds by Pascal and Fermat gave birth to the theory of probability.

This subject had hardly received serious consideration prior to the Renaissance, and had certainly not been calculated mathematically. During the medieval era superstition and fatalism discouraged investigation of the future. This was regarded as the province of God's will, and any attempt at mathematical intrusion here would have been courting a charge of blasphemy.

The mathematical speculation provoked by Pacioli's problem of the points has now become central to our economic life. Upon it rests the calculation of insurance and investment risks, as well as all economic forecasting. How to divide up the winnings (or the likely outcome of a mathematically systemized situation) is now the subject of serious study by experts in every university, commercial institution and racecourse.

In 1497, while Luca Pacioli was lecturing in Milan, he met Leonardo da Vinci. They struck up a firm friendship, so much so that Luca and Leonardo were soon sharing lodgings. Leonardo had long been interested in the geometrical problems of perspective, but this was small beer to Pacioli, and it was he who introduced Leonardo to the hard stuff of real mathematics. Leonardo was initially bamboozled. Here was a challenge the equal of his voracious intellect. He at once began to teach himself this new subject, availing himself of expert tuition from his fellow lodger. No other person was to have such a transforming effect on Leonardo's mind. From now on in his notebooks we see Leonardo's attempts to come to grips with multiplication and fractions. He works out how they exhibit themselves in the proportions of perspective, and moves on to precisely sketched geometric sections of spheres and sliced polyhedrons. He also attempts arithmetical problems. We can imagine the messianically bearded sage (of the celebrated self-portrait) sitting perplexed beside the beefy-faced monk with the still, penetrating eye (of the portrait in Naples) as together they pore over a sheet of figures. At one point in his notebooks Leonardo tells himself that he must 'learn the multiplication of the roots from maestro Luca'. It is heartening to imagine one of the finest mathematicians of his time instructing one of the finest minds of all time in multiplication tables which many of us have now mastered by the age of seven.

Studying the works of Piero della Francesca, Leonardo saw the mathematical principle lying behind the painted appearance. Despite being a mathematician, Pacioli saw a different world: the fluidity of society well before it solidifies into art. The ordering principle, which lay beneath the painted appearances and coloured cheeks which he observed, was money. Mathematics was much more than mere abstraction, more even than the divine proportions of art – it was also the delineation of money.

In his masterpiece, *Summa de Arithmetica, Geometrica, Proportioni et Proportionalita*, Pacioli described all mathematical knowledge. Or at least he attempted to. In those days the attempt to comprehend all human knowledge was still considered a plausible individual aim. Pacioli embraced everything from the pure mathematical knowledge of the ancient Greeks to the latest makeshift measurements of the movements of the heavenly bodies – as astronomy groped its way towards the findings of Copernicus. It was this incorporation of the works of others which led Vasari to say that Pacioli 'gave himself fine feathers'. This judgement, more than any, caused Pacioli to be overlooked by posterity as a mere plagiarist. But Vasari missed the point: Pacioli didn't claim all this knowledge as his own. He was an encyclopedist rather than a plagiarist.

The section in the *Summa* which remains of most interest to us is *Particularis de Computis et Scripturis* (Details of Book-keeping and Ledgers). This contained his explanation of double-entry book-keeping. It may not have been original, but it was the most comprehensive and comprehensible account yet to appear. Indicatively, like the rest of the book, it was written in Italian, the vulgar language of the people, not the Latin of scholars. This was for the use of businessmen and merchants, men not necessarily educated in anything but the subtle ways and time-honoured practices of commerce.

In essence, Pacioli's version of double-entry book-keeping required each transaction to be entered into the ledger twice – as a debit in the left-hand column and a credit in the right-hand column. At any time a line could be drawn under both columns to see if they balanced out, thus revealing any inadvertent or less innocent mistakes in the accounts. Pacioli's method of double-entry book-keeping also facilitated the calculation of the profit or loss in a business at any given time, or over any given period. Such was the marvellous power and utility of this method that its very jargon entered everyday language. Profit and loss, assets and liabilities, balance sheets, debit and credit, bottom line. Amidst the ebb and flow of commerce, the tide of currency could momentarily be frozen into icily precise figures. The process of business could now be subject to mathematical scrutiny and control.

Here, commerce revealed itself as the most advanced science of its day. Mathematics may have begun to play an important role in painting, but physics remained bound by the metaphysical edicts of Aristotle, chemistry was still mired in alchemy and medicine adhered to speculative theories dating from classical times. Not until 1543 would Copernicus discern mathematically a clear concentric picture amongst the complex gyrations of the heavenly bodies. And it would be half a century later before Galileo would understand that mathematics was the key to physics. Commerce had always dealt in figures, but with the mathematics of double-entry book-keeping it fashioned a mirror for itself. It could see its reflection. This self-consciousness was the first step towards reflecting on what it was – what it was for, and how it could be altered.

The spread of double-entry book-keeping through Europe prior to and during the early Renaissance has been seen as the birth of capitalism. The tactile property and possessions of an individual merchant underwent a subtle transformation into abstract assets and money. Goods became value: entries in a ledger which could be used to finance further projects. This also heralded another crucial development in human thinking. The profusion of attributes and qualities which attached to something would tend to be reduced to measurable quantity. And the scale of such quantity would be money. From now on, every thing would have a price.

At the same time European trading also underwent a profound quantative change. In 1492, in search of a trade route to China, Columbus unexpectedly made landfall in America (the world was over a third larger than the experts had predicted). Six years later Vasco da Gama rounded the Cape of Good Hope and reached India, where his beads and cheap cloth were greeted with derision by the Muslim merchants. The Europeans soon learnt to adapt to their new continental markets. The Portuguese took control of the sea trade, and began shipping silk and spices to Europe; while in the Americas, the local inhabitants were simply put to slavery in the mines to dig for silver. The result was an influx of riches such as Europe had never seen before. Meanwhile papal corruption led to the Reformation, with the Continent dividing into Catholic and Protestant states. The stasis of medieval Europe was giving way to social, spiritual and commerical development. When Queen Elizabeth I travelled about England during the 1590s, towards the end of her long and glorious reign, she noticed a profound change had taken place in her kingdom. At every town and village there were clusters of ragged, distressed paupers, many of them homeless. The old certainties of the feudal system were breaking down, and the country was faced with a novel problem: unemployment.

Half a century later England would be plunged into a civil war, and the religion-divided nations of Europe would be in the throes of the ruinous Thirty Years War. This was to leave the heart of Europe devastated, with a poverty-stricken Germany fragmented into a myriad tiny states and princedoms. Vulnerable rulers soon came to realize the benefit of well-organized trade in accruing power at the expense of their neighbour, and began hiring advisors on how to run things.

The ideas put forward by these advisors were seldom original, complex or far-sighted. They had to be understood by a ruler, and had to bring about immediate results – or a new advisor was liable to be appointed. Such ideas usually reflected the practice of successful merchants, and in consequence they came to be known as Mercantilism. Understandably, given the source of these ideas, Mercantilism stressed the importance of foreign trade. Exports should always be greater than imports, thus ensuring a positive balance of trade. Imports should preferably come from distant places, so as not to enrich neighbouring countries – and, of course, rival merchants. Exports created employment, whereas cheap foreign imports destroyed it. These early economic advisors (one can hardly call them theorists) were practical, and were well aware of the difficulties, sufferings and instability wrought in a small state by unemployment. (By contrast, the great theorists who would emerge a century and a half later simply assumed full employment, and thus ignored or failed to see the importance of many Mercantilist recommendations.)

Another Mercantilist virtue was saving. This was to be encouraged, in order to create capital. Capitalism was now passing from the ledger-book into everyday life. A nation's wealth was judged by the amount of gold and silver it had accrued. It is no accident that this emphasis on gold coincided with the growth throughout Europe of towns and cities. Urbanism brought about an increase in the money economy. Money became wealth. Money and gold were equated – and indeed were still largely the same thing. A gold or silver coin worth one ducat was expected to contain gold or silver worth one ducat. Though over the years this equivalence would become increasingly eroded. Coins would frequently weigh less than their equivalent, or be adulterated with lesser metals. But the illusion was maintained, largely for the benefit of the spending public, who wished to believe in the coinage, and would continue to subscribe to this myth of value. To this day, the myth has only been fully exposed and admitted in such countries as Turkey and

Vietnam. Here coins are no longer produced as currency, ironically because the cost of production is more than their face value.

Many Mercantilists were inclined to believe that a nation's wealth could be measured in terms of the amount of gold or silver it possessed. This had a certain validity for the states of Germany, northern Europe and Italy. Here, currency-generating trade was integrally related to productive local industries, such as weaving and mining. Yet for the richest nation in Europe the belief in the intrinsic wealth of gold and silver was to prove disastrous. Spain's New World discoveries included a mile-wide mountain of silver at Potosi, over 10,000 feet up in the Andes in what is now Bolivia. Mining began at once, on a large scale. Inca gold was also appropriated, and all this booty was shipped back in 'treasure galleons'. But the arrival of so much silver and gold in largely rural Spain, where little industry existed, exposed a basic flaw in the notion of gold as wealth. As the supply of gold and silver coins increased, more cash was available to buy the largely unchanging amount of agricultural and other goods which arrived in the market-place – with the inevitable result that prices increased. And went on increasing. In the century following the discovery of the New World, prices in Spain increased by an unprecedented 400 per cent. The newly arrived gold was used to import cheaper goods and grain, to equip the ill-fated Armada against England and to support the army fighting in the Netherlands. The gold and silver flowed in at Cadiz and Seville, and flowed out again at the ports on the north and east coasts where the imports were arriving. Meanwhile in the Netherlands, England, the Baltic and Italy, this same gold and silver was used as capital and 'the sinews of war' (reserve bullion for buying arms). Gold made some richer, but it also made others poorer.

Despite its inherently conservative aim, Mercantilism's aggressive patriotic attitude managed to throw up some colourful characters, and just a few highly imaginative ideas. The attitude is best summed up in the title of a tract by the English Mercantilist

Andrew Yarranton: *England's Improvement by Sea and Land, to Outdo the Dutch without Fighting, to Pay Debts without Moneys, to Set to Work all the Poor of England.* Credit was an economic panacea: it could defeat enemies, balance the budget, even solve unemployment.

The most intriguing of the Mercantilists was Johann Becher, who was born in Germany in 1635. Becher was a man of geniune talents, which he over-extended in all directions with supreme self-confidence. Despite his lack of experience, he managed to persuade his father-in-law to appoint him professor of medicine at Mainz as a wedding gift. But his lasting scientific achievement was in the field of chemistry, which he had come to by way of alchemy. Becher was responsible for the famous phlogiston theory, which was to be accepted throughout Europe for well over a century. According to this, when something burned it released a mysterious substance called phlogiston. There was only one snag. When a thing burns it becomes heavier, appearing to absorb something rather than release it. Adherents of the phlogiston theory re-sorted to the ingenious explanation that phlogiston had negative weight.

Phlogiston theory may be seen as a metaphor for a certain type of economic thinking. Indeed, Becher's Mercantiiism was backed by a similar tenacious ingenuity. Here too, he based his thinking on a well-known fundamental process: what one person spends, another gains. This was the 'soul' of economic life. When expend-iture on consumer goods increased, this had a vivifying effect on all aspects of society, even leading to an increase in the population. So far so good. Becher put these ideas into practice in a number of small princedoms, graduating eventually to the state of Bavaria and then the Austrian Empire. At the same time he also confidently exercised his various talents as court physician, part-time alchem-ist, silk-worm industrialist, inventor of a universal language and whathaveyou – until such time as the ruler in question had had enough, and it was time to move on. Becher's economic ideas were for the most part sound; it was his other talents which usually seem to have let him down. However, it was undeniably his economic theory which led to his eventual downfall.

In 1678 Becher talked himself into a post as advisor to the Dutch government, which was at the time suffering from a gold shortage. Like many Mercantilists, Becher placed great importance on gold as a measure of a nation's wealth, and set out at once to remedy the Dutch shortfall. In a masterstroke he succeeded in combining his talents as economic advisor, alchemist and man of confidence. Appearing before the Dutch assembly, he presented a scheme for turning the sands along the shores of the Netherlands into gold by alchemical means. Astonishingly, the Dutch assembly was convinced, and provided Becher with sufficient funds to go into mass production. At this point Becher's supreme confidence appears to have failed him, for he caught the boat to England. (Or perhaps he had suddenly realized, as the Dutch assembly evidently had not, that gold was far from being the economic panacea it was cracked up to be. Had Becher's scheme actually worked, the Netherlands would have ruined itself as surely as Spain.) In London, Becher quickly regained his confidence, advertising that he had designed a perpetual motion machine. Sadly, this was not a euphemism for his economic theory, whose Merchantilistic ideas had often worked – when unencumbered by such props as alchemy, unlimited gold and unlimited confidence. Becher's 'working model' of his perpetual motion machine was rejected out of hand by the Royal Society, and he died in poverty in London a couple of years later.

Becher's scheme for turning the sands of the Netherlands into gold was not the first fantastic financial scheme in which the Dutch had involved themselves. During the seventeenth century this generally level-headed people became the most financially sophisticated in Europe. They also became the most prone to financial delusion. As we shall see, this combination of financial sophistication and self-delusion has remained a permanent feature – with both individual economists and larger groups of people. An early example of the latter marked the Dutch baptism in financial expertise, and involved the plant for which this nation is now famous: namely, the tulip.

Tulips first arrived in Europe from the Middle East in the sixteenth century. The Viennese ambassador to Turkey is said to have returned with a basketful of tulips bulbs. These delicate blooms with their vivid petals caught the imagination, and they soon became a luxury item for the gardens of the wealthy. Limited supplies ensured their mounting price. In 1562 a cargo of bulbs from Constantinople (modern Istanbul) arrived at Amsterdam, and the European tulip-growing industry was established in the Netherlands. Prices for the brilliant, uniform-hued 'self-coloured' tulips continued to rise steadily, whilst those for the rarer multi-hued 'broken' varieties shot up. The 'broken' varieties were caused by a virus which broke up the solid colours, allowing the pale under-colour to be revealed in florid striations.

By the early years of the seventeenth century prices for broken blooms had reached ridiculous heights. In the French-speaking region of the Netherlands (modern south Belgium) a single bulb was exchanged for a small brewery (*brasserie*), causing this variety to become known as Tulipe Brasserie. Tulipmania had begun, and soon swept through the Netherlands and beyond. In 1611 the new Bourse was completed in Amsterdam, the first recognizable modern stock exchange, and the rising price of tulips was soon being quoted. Prices became so high that innovative techniques in speculation were introduced. For a relatively small sum you could obtain the right to buy a tulip at a future date, at a predeter-mined price. If by that date the price of the tulip had risen higher than this price, profit was assured – without any heavy outlay. This type of trading was called buying an 'option'. What is now known as the futures market was coming into being. Exotic bulbs would frequently change owners a number of times before they had even left the ground. In 1635 a single bulb of the most exotic bloom of

The tulip Semper Augustus, *with its characteristic 'broken' petal colouring. This became the most expensive tulip of all during the Dutch tulipmania of the 1630s*

them all, *Semper Augustus*, changed hands for 3,000 gold florins. (In the same year Rembrandt, at the height of his fame, was paid just half this for his grandiose depiction of Belshazzar's Feast, which with uncanny prophecy depicts the turbaned Babylonian king in all his glory being startled by the writing on the wall.)

'Everything has a price' had now progressed to the point where the price was everything. The intrinsically worthless bloom was now priceless. Money had detached itself from the reality it evaluated. But a morsel of reality did remain – as was discovered by a rich merchant when a sailor visited his house and ate what he thought was an onion, but turned out to be a bulb of *Semper Augustus*.

As the price of tulip bulbs rose still further, so more speculators were enticed to buy in, thus calming the trepidations and reinforcing the expectations of those who had preceded them. Confidence built up from its solid, ever-expanding base like a pyramid, or a house of cards. By 1637 the market in the Netherlands appeared to be all but saturated. Yet dealers knew that by now buyers from all over Europe were flocking to the Netherlands to invest in tulips, and would continue to do so. Or would they? The pan-European Thirty Years War was at its height and the supply of money was beginning to dry up. A few cautious dealers started quietly cashing in their profits. But whispers of these deals quickly spread. Suddenly everyone wanted to sell their bulbs, a panic set in and the market crashed. The price of a tulip collapsed to the price of its functional reality: a simple flower bulb for planting in the garden. Meanwhile options were called in at previously agreed prices, and investors were bankrupted. Some rich burghers were wiped out almost overnight; great merchant houses were ruined; and countless small investors were returned to the poverty from which they had only just succeeded in escaping.

The Dutch tulipmania was the first great speculation bubble in financial history. Such crashes would now become a regular feature of the stock exchange markets. And despite all efforts to regulate against this instability, they would remain so – and will continue to remain so. Instability lies at the heart of the system, and is integral to it. Gambling must always have winners and losers. Speculation can temporarily separate this conjunction. It is possible for everyone to win (as the bubble expands) but this means it is also possible for everyone to lose (when the bubble bursts).

Tulipmania was to have its lasting heritage. In the multi-volume *Palgrave Dictionary of Economics*, regarded by many as the definitive bible of the subject, tulipmania is even listed as a general economic term. 'It refers to a situation in which some prices behave in a way that appears not to be fully explainable by economic "fundamentals".' So tulipmania has become synonymous with an economic wild card, the element of chaos, the flutter of the butterfly wing (or falling tulip petal) which later results in an economic tornado. It has also bequeathed to the Netherlands the most spectacular feature of its flat, wide-horizoned countryside: the brilliant hues of its seemingly endless tulip fields. The vividness of these fields stems from the fact that they mostly contain the mono-hued 'self-coloured' variety of tulip. The 'broken' varieties, such as the fabled *Semper Augustus*, have now for the most part become extinct, eliminated by the very virus which created them. Alas, human avarice has no such improving parable.

But still one more element was necessary before economic thinking could begin to emerge as something more than a hit-and-miss sideshow in the circus of human thought. This element was provided by a modest Englishman possessed of that rare combination of blinkered persistence and wide-ranging imagination. John Graunt was born in London in 1620. According to his contemporary John Aubrey, who included Graunt in his *Brief Lives*, Graunt 'was bred-up (as the fashion was then) in the Puritan way'. Following in his father's trade, he became a haberdasher, opening his own shop, which specialized in accessories such as buttons and lace collars. Graunt had 'an excellent working head' and the business flourished. Yet he also 'rose early in the morning to study before Shop-time'. The novel field Graunt chose to study was to transform our social self-understanding. Graunt copied down details from the City of London Bills of Mortality (death registers). His inital motive appears to have been commercial. He wanted to find out the extent, age and composition of his market. But he soon fell in love with the theoretical and interpretative aspects of the facts he was gathering. In doing so, Graunt became the father of statistics, which would become one of the major tools of economics.

Statistics is based on the systematic gathering of facts. The first such gathering of facts about England occurred in 1085, when William the Conqueror wished to discover the precise extent and composition of the country he had conquered. This was recorded in the Domesday Book – literally, the book of 'doom's day', i.e. the Day of Judgement. The facts gathered were intended as the ultimate authority, as final as judgement day, against which there could be no appeal. Details of every parish and property throughout the land were recorded. Interestingly, however, no mention was made of the people living in these places. Only with the increase of urban living in the sixteenth century did people enter the picture – for the purposes of tax and military service. The first Bills of Mortality for the City of London were collected in 1603, during the bad outbreak of plague in the year Queen Elizabeth I died. Weekly lists were taken of the number of people who had died, with the figures broken down into the different causes of death. The authorities wished to determine whether the plague was on the increase or dying out. From then on this practice continued.

These morbid lists catch the imagination now just as much as they must have caught Graunt's imagination then. In describing death, they bring a city to life. A typical weekly list includes the following reasons for death: 'Murthered' (2), 'Affrighted' (3), 'Suddenly' (2), 'Made away themselves' (2), 'Lethargie' (3), 'Lunatique' (1), 'Found dead (an infant) at St Giles in the Fields' (1), 'Killed by a fall from the bellfry at Allhallowes the Great' (1), 'Aged' (32), 'Wormes' (11), 'Winde' (3), 'Teeth' (33), 'Surfeit' (49), 'Grief' (3). All human life is here.

The Bills of Mortality also included the numbers of all children christened. Graunt took these figures, along with the mortality figures, to calculate the rate of mortality of infants before the age

27

of six. First he selected the categories of death that only included children under the age of six, such as 'Stilborn', 'Overlaid and starved at Nurse', 'infants', and added these together. Then he fell back on his own observations, and calculated accordingly. Half the deaths from smallpox and measles he reckoned to be children under six; likewise just under one-third of the plague deaths. From this he made the 'inference' that 36 per cent of deaths were of children under the age of six. And using the number of baptisms, he was then able to calculate the rate of infant mortality. This was the birth of statistical inference as we know it.

Graunt began studying the 1658 map of London, and soon worked out how he could make further inferences regarding the population. Using his own estimates, he calculated that on average fifty-four households lived in each 100 square yards, and that each household contained on average eight people (a family, including servants and lodgers). From this he calculated that the population of London was 384,000. We have no certain way of knowing the accuracy of this figure, but all other indications suggest that it was more accurate than the contemporary assumption of two million inhabitants!

Graunt continued to pore over his Bills of Mortality early each morning as the cocks crowed and the dawn rose over the spires and Tudor gables of post-Shakespearean London. Diligently he discerned the hard facts behind the city which the Bard had brought to life. He noted that the male birth rate was higher than the female birth rate, but that the male and female population remained roughly equal. From this he inferred that the male mortality rate was higher. Such thinking led Graunt to his most ambitious undertaking. He drew up what he called a Life Table, giving the figures for life expectancy amongst an average group of 100 people. In order to do this he had to make some bold assumptions. As with his previous assumptions (such as the number per household), some of these were inevitably faulty. But it is his daring, and his methodology, which are original. First he took his figure for infant mortality: 36 per cent. He then used similar 'inference' to calculate that 7 per cent of all deaths under the heading 'Aged' were accounted for by the over-seventies. He then calculated the deaths for each decade between, by assuming that during each decade a further ⅜ of the total would die. In this way he gave the figures for how many out of the original 100 born would survive until the ages of 6 (64), 16 (40), 26 (25), 36 (16), 46 (10), 56 (6), 66 (3), 70 (1). Although Graunt's figures must in many cases have been wide of the mark, these were the very first of their kind. (By comparison, nowadays the figure for reaching seventy is approaching 75 per cent in most advanced Western countries.) Where Shakespeare described the ages of man, Graunt calculated them.

Graunt was well aware of the shortcomings of his methods. When he came to publish his findings, he would depreciatingly preface them: 'Here I have, like a silly Scholeboy, coming to say my Lesson to the World (that Peevish, and Tetchie Master) brought a bundle of rods wherewith to be whipt, for every mistake I have committed.' Yet this remark implies that he knew others would follow in his footsteps. Even more importantly, Graunt was also aware of the inaccuracy of the figures he was dealing with. Few sciences have begun with such scepticism concerning their basic data. For instance, Graunt points out that the figures for the 'French-pox' (syphilis) tended to be too low. This was because physicians wished to spare the families the disgrace, so that 'onely hated persons, and such, whose very Noses were eaten off, were reported'. There is another curious omission which Graunt does not mention. Londoners of this period were far from being an abstemious lot where drink was concerned. Many a latter-day Falstaff must have roistered in the taverns of Cheapside. Yet there is no figure for those who died of drink, or even a recognizable euphemism. 'Surfeit' more often than not indicated over-eating, intestinal complications or food-poisoning. Likewise 'gowt' was more likely to be caused by poor diet than Falstaff's beloved 'sack' (from the French *sec*, despite being made of sweet Madeira wine laced with lime juice). So what became of the boozers? What accounted for this ever-popular form of mortality? Even the earliest statistics still contain their secrets: they are not as dry as they would have us believe.

Graunt's sampling may on occasion have been suspect, but his method transformed the theory of probability. Less than a decade previously, Fermat and Pascal had become fascinated with the calculation of probable winnings resulting from Pacioli's unfinished game of balla. Fermat and Pascal worked out more mathematically accurate formulae for probability, but it was Graunt who extended it beyond mathematics and games of chance into the real world.

In 1662, at the age of forty-two, Graunt published *Natural and Political Observations made upon Bills of Mortality*. In it he includes all his 'inferences' from the City of London death registries, covering the period from their first publication in 1604 until the latest in 1661. This work is generally recognized as the founding of demography, the statistical study of population. Graunt didn't actually use the word statistics, which only came into use in its modern sense a century later. It derives from the Latin word *status*, and is a happy blending of both its meanings: 'nation' and 'condition'. Graunt's work was to transform people's conception of the world in which they lived. And this effect was not limited to scientists and scholars. Throughout the population of seventeenth-century London there was a widespread terror of being struck by lightning. Graunt reassured people by showing them that they were twice as likely to die 'Bit with a mad Dog' (2).

Graunt became friends with the polymath and thinker Sir William Petty, who shared many of Graunt's interests. It is said that Petty assisted Graunt in coming to some of his more far-reaching conclusions. Their contemporary Aubrey goes even further, suggesting of Graunt and his *Observation*s, 'I believe, and partly know, that he had his Hint from his intimate and familiar friend Sir

William Petty.' This insinuation that Petty was the inspiration behind, or was even responsible for, Graunt's work is best dismissed as gossip. Petty himself was influenced by Graunt, and even then he did not comprehend the unique significance of what Graunt was doing. As we have seen, Graunt himself did have intimations of his own immortality – the 'Scholeboy' felt sure he would start a school of thought – yet it is fair to say that he probably didn't realize the full import of what he was starting, or where it would lead. (As Benjamin Disraeli would famously point out two centuries later, 'There are three kinds of lies: lies, damned lies, and statistics'.) When Petty introduced the shopkeeper Graunt to his upper-class scientific friends, he tended to be condescending towards the mere fact-gatherer, rather than suggest that Graunt's work was his own. Aubrey's innuendo is on a par with the snobbish suggestion that Sir Francis Bacon must have written the works of Shakespeare. Such supreme literature could not possibly have been produced by an insignificant actor, the son of an impecunious Stratford glover – a rumour which was gaining currency at just this time.

Graunt's work caused a stir, and not just amongst the lightning-affrighted classes. Charles II was so impressed by *Observations* that he suggested its author should be elected to the prestigious Royal Society, whose members at the time included such luminaries as Sir Christopher Wren, the future architect of St Paul's cathedral, and Sir Robert Boyle, the founder of modern chemistry. When the members of the Royal Society demurred at electing a shopkeeper, the King dismissed their objections: 'If they find any more such Tradesmen, they should be sure to admit them all, without any more ado.'

Yet election to the Royal Society was to be John Graunt's final achievement. From then on his world was to prove very much a 'Peevish, and Tetchie Master'. Graunt ill-advisedly converted to Catholicism, just when popular prejudice was turning once more against 'Papists'. In 1666 he was accused of having 'some Hand'

31

in the Great Fire of London. He had been appointed an officer of a water company, and was said to have cut off the water supply just before the fire was started (allegedly by Papists). A likely story – seeing as his own shop was burnt to the ground, a blow that resulted in his bankruptcy. Graunt died, disgraced and impoverished, nine years later at the age of fifty-three. It would take another 250 years before the pioneering nature of his *Observations* was fully appreciated, and Graunt was recognized as the father of statistics. Over the ensuing centuries, statistics would provide the factual foundations on which economic theory could build. Without statistics, economics would be little more than guesswork.

Aubrey may well have learnt of Sir William Petty's 'role' in Graunt's work from Sir William himself. They were friends, and a number of Sir William's tall tales were recorded by Aubrey. Petty claimed to have taken a ship 'with little stock' to Normandy at the age of fifteen. Here he 'began to merchandize' so successfully that he earned enough to educate himself. Despite not being a Catholic he achieved entry to the celebrated Jesuit college at La Flèche (where Descartes had recently been educated). He then went to Paris, where he allegedly read anatomy with the political philosopher Thomas Hobbes, had conversations with Descartes, exchanged ideas with the mathematician Gassendi and was influenced by meeting the philosopher-priest Mersenne. Yet at the same time he was so poor that 'he lived a Weeke on two peniworth of Walnutts'. Such exaggerations are unnecessary, for Sir William Petty was undoubtedly an exceptional man, worthy of his place amongst his exceptional English intellectual contemporaries. Apart from those already mentioned, these ranged from Harvey, who discovered the circulation of the blood, to Locke and Newton. This was one of those rare periods when England's intellectual transcendence was recognized throughout Europe. Sir William Petty's achievement was in pioneering economic thought, where he had no peer. And like Graunt, who was a key influence, his fundamental contribution was also long unacknowledged.

Sir William Petty was born near the Hampshire coast in 1623. His father too was a haberdasher; and like his friend Graunt, Petty would also later be successful in this trade. It is this, rather than their esoteric researches, which was almost certainly the initial reason for their contact. Despite all the exaggerations by Aubrey, Sir William Petty really was an outstanding businessman. His abilities in this field appear to have been accompanied by a similar unscrupulousness to that which he would exhibit in the intellectual field. As a result, he was to become very rich indeed.

As a youth, Petty may well have run away to sea, but he abandoned the life of a cabin boy at fifteen and probably took over his father's shop. It was here that he earned the money for his education, which included studying medicine at Leiden and Paris, ending up at Oxford, where he became professor of anatomy at the age of twenty-eight. This post bespeaks exceptional ability on Petty's behalf – and not only on account of his comparative youth. Petty was also extremely short-sighted, to the point where he almost couldn't see beyond his nose (an affliction which can't exactly have inspired confidence in his patients on the operating table). A year later Petty accepted the chair of music (sic) at Gresham College in London. How long this lasted isn't quite clear. The prospect of achieving wealth and influence in the capital appears to have been his main consideration. By the end of the year he had secured an appointment as chief medical officer to the army in Ireland. Petty has been characterized as a 'bumptious and somewhat unpleasant man'. Indubitably brilliant, all but blind, an extremely sharp businessman, he must also have possessed considerable powers of persuasion. Despite his incapacitating short-sightedness, he now talked his way into the leadership of a topographical expedition which was to map the whole of Ireland. The aim appears to have been to produce at the same time a sort of Irish Domesday Book. The survey itself was conducted with the utmost diligence and scrupulousness. Not one parcel of land in the entire island went unaccounted for: its entitled owner was noted, its worth recorded. According to the current practice, this was fixed at twenty times the annual income it produced from farming or rent. What was to become of those lands with disputed or obscure ownership was less clear. What is clear is that three years later, after the completion of his survey, Sir William Petty emerged with estates throughout Ireland. According to Aubrey he could 'from the Mount Mangorton in ... Kerry behold 50,000 acres of his owne land'.

This new-found wealth did little to increase Petty's popularity. It comes as no surprise to learn that in 1660 he was challenged by an irate Irish landowner, Sir Aleyn Brodrick, to a duel. Petty's response gives some measure of the man. Despite his physical disability, to ignore the challenge would have exposed him to censure as a coward and public mockery. So he exercised his right, as the receiver of the challenge, to name both the weapons and the location of the duel. He chose 'a dark cellar, and the weapon to be a great carpenter's axe'. This turned the whole thing into a farce, and Brodrick withdrew in disgust.

Petty spent his next years administering his estates, living off his vast rents, and vigorously contesting all challenges to his ownership in the courts. He became a member of parliament, and also applied his mind to all manner of intellectual projects and inventions. The latter included an 'unsinkable' double-hulled boat, built 250 years before the *Titanic*, which ended up in the same place as its mighty successor. Yet these activities were little more than the recreation of an ingenious and fertile intellect. Petty's prevailing interest, which exercised his genius to its full powers, was economic theory. In many ways, he was the first consummate thinker in this field, and the ideas he set down were to provide inspiration (to put it at its narrowest) for the man more usually recognized as the first economic theorist in the classic canon: Adam Smith.

For all its originality, Petty's thought was in many ways a product of its age. His outlook was empirical and scientific. The same approach which would lead Newton to explain the world in terms of gravity inspired Petty to explain the economic world in terms of what he called his 'political arithmetick'. The world of economics, like the world of astronomy, was to be mathematical.

Petty's thought was also a product of his personal circumstances and character. He made full use of the experience he had gained in running his Irish estates, and his temperamental affinity to the ideas of Machiavelli meant that he espoused forthright action that succeeded, regardless of moral considerations. The result was a curious mixture of brutal realism and a compassionate perception of what economics was actually about. His no-nonsense seventeenth-century forthrightness of aim is appealing to the modern scientific sensibility.

> To express myself in Terms of Number, Weight or Measure; to use only Arguments of Sense [i.e. Reason], and to consider only such Causes, as have Visible Foundations in Nature; leaving those that depend upon the mutable Minds, Opinions, Appetites and Passions of particular Men, to the Consideration of others.

But this same frankness can also be grating, at least to this author's Irish sensibility: 'As Students of Medicine practice their inquiries upon cheap and common Animals . . . I have chosen Ireland as such a Political Animal.' Yet as we shall see, Petty's cold inner eye was not short-sighted.

Petty wrote his *Essays in Political Arithmetick and Political Survey or Anatomy of Ireland* in 1672. In this he set out 'to make a Par and Equation between Lands and Labour so as to express the value of any thing by either alone'. This he did by calculating the value of labour in the same way as the value of land. A man was to be valued at twenty times his annual income. In this way it was possible to calculate the loss to a nation by deaths, especially in times of war and plague. But this was only the beginning.

Just fifty years previously Galileo had declared: 'the Universe ... is written in mathematical characters'. Petty now sought to extend this scientific belief from the physical universe to the social universe. The fields, farms and towns of a country, as well as its men, women and children, could now be reduced to a common figure: their monetary value. In fact, Petty was not so much interested in money for its own sake. The value of money only lay in what it valued, what it could purchase, what could be exchanged for it. Petty saw money more as a common denominator against which everything could be measured, so that equivalences could be established across the board.

Yet if the countryside and its people could be reduced to figures, so too could everything else. Petty foresaw a day when a nation's power, its art, and even its opinions could be calculated by 'political arithmetick'. Likewise on the individual scale. Eloquence, reputation and authority would all be measured. This may seem extreme when stated so baldly, but it was very much a reflection of current rationalistic optimism. The German philosopher Leibniz, who had just constructed an early calculating machine, foresaw a future for these machines which extended even beyond the extensive domain of modern computers. Leibniz felt sure that one day all moral and legal disputes would be settled by calculating machines. Feed in the arguments for both sides, the machine would then calculate and deliver its verdict. This assumes a world in which everything is susceptible to measure. We may not yet have reached this stage, but the notion that everything, and everyone, has a price is hardly alien to our thinking. Similarly, opinions are now measured by polls, and emotions are assessed scientifically by psychology.

Petty was searching for the laws which underlay the workings of a nation. If money was the measure of society, then it could be used to control it. Money could be the instrument for the effective running of the country. Here, his Machiavellian instincts couldn't resist pointing out the shocking implication of all this. If a country could be controlled by money, then it would have no need of kings or priests. Money, it seems, could exercise power and promote moral qualities. It is important to stress that Petty did not view money as the be-all and end-all of social life. He saw money as the measure, the method of controlling the social system. It is the manipulation of this measure which exercises control – bringing efficacy and happiness, or the opposite.

Prior to Newton's discovery of gravity, another important seventeenth-century scientific advance had been Harvey's discovery of the circulation of the blood. As a former professor of anatomy, Petty understood this process intimately. He saw money as fulfilling a similar role in the body politic. Its circulation was essential to the life of society; the body politic too needed this flow to maintain its social organs and limbs. Continuing this analogy further, he insisted that no part of the body could simply be ignored and left to rot, without inflicting severe damage upon the whole. If money was to flow throughout the system, this meant that the state had a duty to maintain as high a level of employment as possible. If unemployment could not be prevented by means of money, then the state should provide public works projects, such as the building of new roads. This kept labourers employed, and also contributed to the nation itself. Petty could demonstrate this by reverting to his previous analysis of income and rent, and their equivalence. The unemployed labourers were given an income, and were thus worth more; the road provided more rent in the form of turnpikes and tolls, and thus it too was worth more. All this increased the nation's wealth.

The Mercantilists had reflected the virtues of successful trading. Saving was good moral as well as good economic practice. Petty reflected the mechanistic world of the new science. Measures were to be taken because they worked, not because they were good (here too we can once again see Machiavelli's influence).

37

Petty's cold, amoral viewpoint, allied to his exceptional insight, also enabled him to make some uncannily accurate predictions. One of these is known to this day as Petty's Law. This states that as an economy develops, the proportion of the working population employed in services will tend to increase. It is worth remembering that Petty discerned this trend whilst running his country estates in Ireland. This law, which still holds true in the world of silicon valley and the 'new economy', was deduced from a working population of peasants, where the service industry was liable to consist of 'greasy Nell doth keel the pot' up at the manor house.

Another prescient insight came with Petty's analysis of surplus. Where land was concerned, this could be viewed as the quantity of grain produced over and above the cost of its production (labour, seed etc.). This fell into the same category as the rent which could be gained from a piece of land. In other words, twenty times the annual surplus was its value. This also realistically reflected the rise and fall of the price of land in times of plenty and scarcity. Though increased harvests could not raise the price of land in direct proportion, because the price of grain would inevitably fall during abundant harvests.

Petty did indicate another way in which surplus could be measured. It could be seen as the number of unemployed people who can be maintained by a company of labourers whose produce is enough to support both groups at subsistence level. Such an insight, and the harsh conditions it evokes, are the very stuff of eighteenth-century Ireland. Only a landlord concerned for his profit, but afraid of the consequences of a starving rabble, would have seen things this way. Yet the same principle continues to apply in the modern welfare state. Petty's aim was to discover the natural laws which governed the anatomy and function of the body politic. He saw himself as searching for the inner structure of social reality. Such structure does not change, even though our attitude towards social reality may have become more compassionate.

In 1682 Petty sketched out his monetary theory in his modestly titled *Quantulumcunque Concerning Money* ('something, be it ever so small, about money'). Petty viewed money as a measure, a repository of wealth and a means of exchange. However, his idea of its function in the body politic suggests a somewhat muddled view of anatomy – especially coming from a professor in this subject. As we have seen, he viewed money like blood: its circulation kept the limbs and social organs of the body politic alive. But he now states, 'Money is but the Fat of the Body-politick, whereof too much does often hinder its agility.' Such confusion might be fatal on the operating table, but it made economic sense. Petty understood that a nation's wealth did not lie in the amount of money it possessed. 'The Blood and nutritive Juices of the Body politick [are] the product of husbandry and manufacture.' He saw full well what had been the ruin of Spain. This downgrading of money's supreme importance worked on the individual level as well. If labourers were paid too much money, labour was 'scarce to be had at all, so licentious are they who labour only to eat, or rather drink'. In other words, when people earned enough money to enjoy themselves, they simply stopped work and did just that. This may appear a typical landowner's view of his labourers, and it is certainly not true nowadays of workers in the Western world (not entirely, at any rate). However, this practice remains prevalent amongst labourers in non-industrial countries in the Third World. What else would a coffee plantation worker in rural Brazil be expected to do with his excess money? The idea that people are motivated to make more money and better themselves could not be assumed as a primary economic motive until the coming of the Industrial Revolution in the following century. Petty rightly saw that in his time the basic principle of individual economic life was much more vague. It could only be loosely termed as self-interest. This could be likened to an inverted form of Newton's gravity. Each individual strove for himself, and in doing so produced a generalized upward impulse in society. Yet this impulse was more to remain afloat, rather than reach for the skies.

39

However, such vague psychological notions were essentially foreign to Petty's central idea. His political arithmetic required exactitude: 'number, weight and measure'. This caused Petty to disregard the wayward variety of individual human nature in favour of an abstract generalization which was measurable. Instead of human beings, he dealt in 'labour'. This was a perhaps unavoidable simplification. It was necessary if economic thinking was to proceed in a scientific manner. The need was for laws, rather than the intuitions of art (compare Petty's 'Fewness of people is real poverty' with the statement of the playwright George Farquhar, who witnessed the same Ireland: 'There's no scandal like rags, nor any crime so shameful as poverty'). Yet it is salutory to remember that, in a genuine science, one instance contradicting a law disproves that law. In economics, on the other hand, the exception tends to prove the rule. The laws of economic theory are statistical, and deal with a statistical entity. In Petty's case this entity was 'labour'. Later sketches of *homo economicus* would differ, tending to a sharper profile. But this entity would always remain essentially a caricature – a lumpen mass of statistics, rising at best to the faceless, politically incorrect 'everyman', or the ever-open-mouthed 'consumer'. Economics may describe the behaviour of a caricature, but it is at least a human caricature. The so-called 'genuine' sciences, such as physics or chemistry, aim for the most part to dispense altogether with the human presence.

Becher had advised rulers on limited policy matters, Petty had tested his theories on his Irish estates, but so far no economic thinker had been granted the freedom to run an entire nation. The opportunity to put theory into practice on the grand scale would not arise until the following century. This would be the ultimate test of the new science, and would produce spectacular results.

2

The Richest Man in the World

The first economic thinker to run an entire country was the Scotsman John Law, who took over in France during the second decade of the eighteenth century. His actions would transform the fortunes of Europe's largest nation. They would also transform his personal fortune beyond the wildest dreams of avarice.

By 1720 John Law was the richest man on earth. Some even claim that he was the richest man in history. This is no idle claim. Law's possessions included the French central bank, the Banque Royale, the only bank in France at the time. Through the Banque Royale he held the entire French Louisiana territory, which stretched from the Gulf of Mexico to the Great Lakes, from the Appalachians through the Midwest to the Rockies (the equivalent of almost two-thirds of the present USA). His companies had a monopoly on French trade with the Americas, including the lucrative slave trade with the West Indies, as well as complete control of French trade with its colonies in India and the Far East. In recognition of his American possessions, Law was made the Duc d'Arkansas. In France he owned over a dozen grand châteaux, and as the owner of many extensive estates throughout the country he had also accumulated a slew of other noble titles. In Paris he owned the Hôtel Nevers, several other grand mansions, as well as an entire block of streets.

Law lived in some style with his English aristocratic mistress, Lady Catherine Knollys, and their two children at the palatial Hôtel de Soissons. Princes and dukes clamoured for invitations to dine with him; duchesses gratefully curtsied before him, kissing his hand. In the evening a greater number of fashionable coaches were to be seen drawn up outside the Hôtel de Soissons than outside the Regent's residence at the Palais Royale. The papal nuncio came to tea and would 'play at dolls' with Law's daughter, evidently prepared to overlook the fact that she was illegitimate and her father was living in sin. According to Law's earliest biographer, J. P. Wood, who encountered contemporaries of Law, 'Few people equalled him in number, splendour, and value of jewels, plate and equipages, though at the same time he took care that the strictest order and propriety should be observed in his household.' He gave generously to charity. Law had but two vices: womanizing and gambling. Managing his extensive business affairs, and administering the finances of the most powerful nation in Europe, left him limited time for the former. As we shall see, however, his financial dealings offered him ample range for the latter.

Law ran his business from a salon in the Hôtel de Soissons. According to his contemporary the Duc de Saint-Simon,

> Officers of the Army and Navy, ladies of title and fashion, and everyone to whom hereditary rank or public employ gave a claim for precedence, were to be found waiting in his antechambers . . . peers, whose dignity would have been outraged if the Regent had made them wait half an hour for an interview, were content to wait six hours.

All wished to do business with the phenomenally successful Scotsman.

But this was only half the story. Unlike almost all plutocrats, Law also remained highly popular with the French people. When he was recognized in his carriage, the populace would break into spontaneous cheers and cries of '*Vive m'sieur Law!*' Law's success had made many of them rich, and their gratitude was heartfelt. There was certainly no resentment at street level towards this foreigner who wielded unprecedented power in France.

An indication of this power is seen in his dealings with the French Regent, Philippe II, the Duc d'Orléans, who went out of his way to assist Law in his business projects. Law was even granted a licence by the French Regent to print money (though this proverbial facility was in fact Law's own suggestion). Not for nothing has John Law become known as 'the father of paper money'; he has even been called 'the founder of our monetary system'. In the light of such claims, a closer examination of Law's life and financial ideas should prove economically instructive.

John Law was born in Edinburgh in 1671, the son of a goldsmith. As was customary in this trade, his father also provided simple banking services, taking cash on deposit and providing loans. The notes given by these goldsmiths to depositors in fact became the earliest form of paper money in general use in the British Isles, though they were not legal tender. Business prospered, and Law Senior was able to buy himself the modest estate of Lauriston along the coast just west of the city. At school his son John showed exceptional talent at arithmetic, and during the holidays he absorbed the requisites of banking. The combination of mathematics and money fascinated him, and he soon began forming ideas of his own about this potent mix.

By the age of twenty-one John Law was living it up in London high society, with no visible source of income to support his expensive tastes in clothes and women. His abstract mathematical-monetary speculations had been put to more practical use in the form of astute gambling, accompanied by quicksilver calculations of the odds involved. But this fast living and quick thinking suddenly took an unforeseen turn. This would blight his life from then on, yet at the same time lift it into a realm where he would be driven to develop his talents to the full. In 1694 John Law became involved in a dispute over a woman, and was challenged to a duel by a notorious rake known as Beau Wilson. During the duel Wilson received a sword 'to the belly' and died soon afterwards. Law was charged with murder, tried and sentenced to be hanged at Tyburn.

Law's duel with Wilson had been the culmination of a complex intrigue involving Elizabeth Villiers, Charles II's mistress, who wanted Wilson out of the way. With the contrivance of Villiers' friends, Law was spirited from prison to await a boat for the Netherlands. A wanted notice, offering a reward, appeared in the *London Gazette*: 'Captain John Lawe, a Scotchman, lately a Prisoner in the King's Bench for Murther, aged 26, a very tall lean Man, well-shaped, above Six foot high, large Pockholes in his Face, big high Nosed, speaks broad and loud.' This description is misleading (intentionally so, one assumes). Law had no military rank, no pock-marked face, and spoke with a restrained gentlemanly accent – though he was tall, 'well-shaped' and had a prominent nose. He was said to have been extremely good-looking, possessed of a personable character which men enjoyed and women found particularly attractive.

Having served his apprenticeship in London, Law was now forced to live by his wits in exile in Europe. The details of his travels during the next decade are as imprecise as Law presumably wished them to remain. He appears to have passed through the society gaming tables of Brussels, Geneva, Genoa and Venice, moving on when the pressing attentions of angry husbands and disgruntled gambling companions became too great. But Law was more than a mere forerunner of Casanova. From now on, his life became increasingly calculated. He played the gentleman down on his luck, with more than a hint of a scandalous past. The ladies were intrigued, and his gambling companions were lulled. But his countless seductions were probably more repute than actuality; and his uncanny ability to calculate odds was used only at decisive moments, so that it might appear to be luck. He acquired a considerable amount of money, yet he wasn't particularly interested in flaunting it. On the contrary, his discreet generosity seemed to some almost selfless. This was a misapprehension. It wasn't the man himself who was selfless, it was his money. Law's way of life was enabling him to gain a profound understanding of what precisely money was. Money didn't have a self, it had a function. Money wasn't pieces of silver or gold, or even the things for which they could be exchanged. It wasn't a thing at all, it was an action. Money unused was nothing – nothing but potential for action. Money should be regarded as a verb, not a noun.

In 1702, while in Paris, Law encountered an aristocratic English-woman called Lady Catherine Seigneur. She was descended from the family of Anne Boleyn, and was married to a Frenchman. By coincidence Catherine's brother had also been imprisoned in England for killing a man in a duel. But John and Catherine soon found they had more in common than exiledom and mere coincidence. Her unhappy marriage, and the constant wariness involved in his shallow role-playing, had turned each of them into inner exiles. For the first time in years Catherine was able to express her feelings. Perhaps for the first time ever, John Law found himself doing the same. On the spur of the moment, they eloped to Italy. Catherine reverted to her maiden name Knollys, and the two of them lived together as man and wife.

Amsterdam was the financial centre of Europe at the time. The Dutch navy held sway in the North Sea, and the Dutch East India Company was the major player in European trade with the East. It was only natural that Law should eventually gravitate to the Netherlands. What he saw there opened his eyes to commercial reality. If money was action, investment in overseas trade was evidently the most effective form of that action. The Netherlands was awash with money and money-making schemes. The powerful Bank of Amsterdam had even introduced a new scheme to put more of this potential action into circulation. As well as accepting deposits of money, it also took over landed property. It would then issue loans in the form of notes secured by this property. As well as a money bank, it had also become a land bank. (As J. K. Galbraith has astutely pointed out, 'Just how the land would be redeemed by noteholders was uncertain.') Law quickly understood the benefits of the Dutch approach. More cash generated more commercial activity. In 1703 Law determined to put his ideas into action, and returned home to Scotland. (At this time England and Scotland remained separate countries: the murder warrant in England did not apply north of the border.)

Scotland was in a parlous state, still reeling from the catastrophic effects of the Darien Venture, which had come to grief five years previously. This was Scotland's attempt to reap the benefits of becoming a colonial power. A transporting route was to be established across the Isthmus of Panama, enabling South Seas trade to link with the Atlantic Ocean. The Venture was led by William Paterson, a Scot of proven financial acumen who had assisted in founding the Bank of England (at the very time when Law had been languishing in jail down the road). 'Trade will increase trade and money will beget money,' Paterson assured enthusiastic investors in the Venture. In 1698 Paterson embarked for Darien in Central America, in charge of a fleet of five heavily laden ships, which represented half the capital in Scotland. A settlement called New Caledonia was duly established on the coast of the Isthmus of Panama. The surrounding land turned out to consist of impenetrable jungle and disease-infested swamp, the settlement was fiercely contested by the local Spanish settlers, and within seven months the new colonists set sail back for Scotland. Just 300 of the original 1,200 colonists had survived, and Scottish capital worth £400,000 had been wiped out. Meanwhile, unaware of the fate of their predecessors, two further expeditions set out for Darien, to meet a similar fate and provide a further drain on the nation's meagre finances. The Scots, traditionally associated with financial prudence, were devastated. According to Scottish historian John Prebble, this was 'perhaps the worst disaster in Scotland's history'.

William Law arrived back in Scotland convinced that he knew how to repair the damage. He decided to set down his ideas, which he duly published in 1705 as *Money and Trade Considered with a Proposal for Supplying the Nation with Money*. The operative word appears twice in the title. Law opens by describing the history, not of money, but of its use. To be of any value, money must be used, it must circulate from hand to hand. When there is not enough money, credit notes will appear. In such times, people who work or provide services may even be paid in credit notes. Thus, these credit notes too become a form of money. He goes on to explain that when a nation is short of money, one remedy is to mine more gold or silver and mint more coinage. But what if this nation has no such mines, like Scotland? The use of credit notes showed the way. When more money was required you didn't need a gold mine, you needed a bank. This could issue credit notes against the one quantity which Scotland possesed in abundance – land.

Law placed his proposals before the Scottish parliament. He was aware that such a radical idea was liable to provoke opposition, but he was capable of arguing his case with considerable skill. As *Money and Trade* demonstrated, Law had thought long and hard about money and what it meant. And his Continent-wide experience of how it could be accumulated, both by himself and others, spoke for itself. But the opposition Law in fact incurred came from an unexpected quarter. He was accused of plagiarism by an Englishman called Dr Chamberlen, who had once been personal physician to Charles II. In 1695 Chamberlen had set up a land bank in London. This had been a covert imitation of the Bank of Amsterdam, though he had claimed the idea as his own. Either way, four years later Chamberlen had found it expedient to catch a night sailing on the Amsterdam packet, leaving the clients of his bank with nothing but worthless pieces of paper (in the event, Galbraith's qualms proved justified). Chamberlen too had taken advantage of the fact that English law did not apply in Scotland, and was now resident in Edinburgh. Chamberlen had powerful friends, so Law decided to drop the idea of a land bank. Instead he proposed an even more revolutionary idea. Rather than issuing credit notes against land, the government should issue them against itself! This was not such a legerdemain as it might appear: the government would simply be issuing credit against its own

47

ability to raise money in the future, such as by taxation. This of course depended upon the nation's commercial performance.

Such money is now known as fiat money, and appears by a fiduciary issue of notes not backed by gold. The word fiduciary derives from the Latin *fiducia*, which means 'confidence'. Fiduciary was first used in this context by the Bank of England in 1844, because it was felt that where money was concerned the English version had less reassuring associations – though such issues of banknotes remain, of course, a confidence trick in the literal sense. There must be confidence in the future commercial performance of the nation. Here Law's suggestion was well ahead of its time. Nowadays all our paper money is a fiduciary trick.

Law's ingenious suggestion was rejected, and two years later parliament signed the Act of Union with England. Whereupon the nation lost two of its most prominent financial advisors, when both Dr Chamberlen and John Law chose to retire once more to the Continent. Here, Law returned to his old profession, which he continued to pursue at the tables in Paris and then Turin. Besides Catherine Knollys, he now also had a son and daughter to support, in the manner required of a gentleman gambler. However, he also remained interested in more abstract monetary matters, and continued to champion his ingenious financial idea. In Turin, this eventually came to the notice of Duke Victor Amadeus of Savoy, who was so impressed that he determined to put it into practice. But the conservative Savoyard bankers, who saw Turin as a future European financial centre, were adamantly opposed to such a controversial innovation.

Victor Amadeus suggested that instead Law should propose his idea to his brother-in-law, Philippe II, the Duc d'Orléans, who had recently become Regent of France. Law set off for Paris with a ducal recommendation in his pocket. Here was his big chance.

In 1715 France was the largest and most powerful country in Europe. Yet its finances were in a catastrophic state, following the death of Louis XIV, 'the Sun King', earlier in the year. Louis' long reign had been characterized by constant costly wars and large-scale regal extravagance. The latter had reached its apogee in the building of Versailles, the largest palace the world had yet seen, home to a court of 20,000. The treasury had been left empty and beset by colossal debts.

Louis XIV had been succeeded by his ailing five-year-old great-grandson, under the regency of Philippe II. The new Regent was a man of cultivated taste and uncultivated intellect, who preferred to conduct his extravagances in sophisticated Paris rather than out at Versailles. Paris was at the time the most cultured city in Europe, its theatre and concerts unrivalled, the ideas of Voltaire and Montesquieu were beginning to surface in the salons. Philippe moved the court to the Palais Royale, where he proceeded to hold grand masked balls to show off his mistresses, and indulge in private orgies with his roué friends. (The word 'roué' was in fact coined to describe Philippe's coterie. It was short for *digne de roué*, meaning 'fit for the wheel', a reference to the old French torture of being broken on the wheel.)

This is the man John Law now approached with his new financial idea, which he presented with the self-confidence and winning manner he had used to charm his way around the capitals of Europe. He knew that the French exchequer was faced with debts of around three billion *livres tournois* (at a time when anyone with a private fortune of one million *livres* was rich). Despite this, Law promised Philippe, 'I will devise a scheme that will astonish Europe by the advance it will make in France's favour, alterations more radical than those procured by the discovery of the Indies or the introduction of credit.' He first proposed the setting up of a bank, which would be followed by a scheme 'which would raise 500 millions without cost to the people'. This miraculous transformation of the nation's finances would make France 'the arbiter of Europe without recourse to violence'.

Philippe and Law quickly found they had much in common. The Regent was captivated by both the man and his scheme. (And just to cement matters, it is alleged that Law made a conquest of Philippe's widowed mother, the ageing but influential German princess Elisabeth Charlotte d'Orléans.) This foreigner, who supported himself as a professional gambler and womanizer, and was also an escaped murderer, now became the Regent's confidant and financial advisor. Two crucial factors worked heavily in Law's favour. Catherine Knollys had taught him to speak aristocratic French without even a trace of an accent. This made him appear to be more than simply a visiting Scotsman, and helped overcome much customary xenophobia. Also, France remained backward in financial matters. Compared with liberal Britain or the Netherlands, which were already beginning to develop new banking methods and complex commercial enterprises, autocratic France remained a centralized, largely rural economy. France was in dire need of new financial ideas, and Law was the man who had them. He had never lacked confidence, but he had never previously had any practical experience whatsoever in the field. This would soon be remedied.

In May 1716 Law was given dispensation to open a bank, the first in France. This bank, which would eventually be known as the Banque Royale, was granted a capital of 6 million *livres*. The bank duly opened for business on the first floor of Law's impressive house in the fashionable Place Louis-le-Grand. Here, cash deposits, in the form of coins, were accepted on account; and a system much like modern cheques was put into action so that individual account holders could make transfers between each other's accounts. In fact, it ran in much the same way as a modern high-street deposit bank. However, as this was the first of its kind in France, it had almost none of the safeguards which have been built into the banking system over the years. This was hardly surprising. Such safeguards only came about with hindsight, as certain flaws in this mode of operation came to light – or were exposed by those gifted with foresight. As with any pioneer commercial enterprise, the risk was great (and the possibility of gain was equally considerable).

Yet deposits and cheques were far from being the main feature of France's first bank. Its *raison d'être* was of a different order altogether. Law's innovative suggestion was that his bank should also be authorized to issue paper money. This would be used to cover the government expenses which could not be met by the exchequer in the form of 'real' cash. The issue of these 'bank notes' was covered by the bank's original 6 million *livres* capital; and the notes could, if the bearer so wished, be exchanged for coins at the bank. At least, this was the theory. In fact, the bank simply didn't have 6 million *livres* in cash. Its capital consisted largely of government promissory notes and state bonds, with just 350,000 *livres* in actual coin.

Law wasn't too worried about this. There is no doubting that he had belief in his ideas. And curiously, it quickly becomes clear that he had not set up this scheme in order to make money for himself. What interested him most of all was to see if his ideas worked. And because they were ideas of genius, they did. (As we have seen, it is upon such ideas that our present financial system rests. Which means that at this point we should all have a very keen interest in seeing exactly how these ideas fared.)

The banknotes issued by Law's bank were a success from the word go, as were the deposit accounts and cheques. Money was loosened up. It no longer weighed you down, it was more easily hidden or carried. Money could be moved around more easily. And just as Law had foreseen, this issue of banknotes was also a success in the larger scheme of things. Simply creating money out of nothing (the banknotes used to pay off government debts) had a galvanizing effect on French society. People now had more money, and spent it. This created a demand for more goods, and France quickly began to rouse from its pastoral slumbers as more people made these goods, and made more money doing so. Indeed, these handy new banknotes proved so popular that people began bringing coins to Law's bank and demanding notes, so that he was forced to print more. Then he had to bring out a further issue of

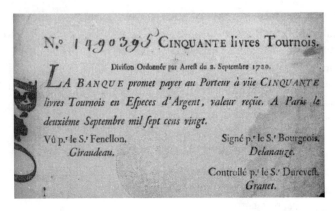

Note for 50 livres tournois *issued by John Law's Banque Royale*

notes to cover further goverment debts which came to light.

But the coins coming into the bank were a trickle compared to the positive deluge of notes with which France was soon awash. By 1717, in order to maintain confidence in the new banknotes, they had even become legal tender for the payment of taxes. Much as Law had foreseen, what he now needed was a big inflow of hard cash. This he could then use to buy land, as collateral against the bank's paper debts, at the same time recirculating the hard cash so that it could continue vitalizing French commercial life. The Bank Royale moved into the Hôtel Nevers, an impressive mansion beside the Bibliothèque Nationale, the French national library.

With the Banque Royale very much a going concern, Law turned to 'the scheme' to which he had alluded in his original proposal to Philippe. He had promised that this would 'astonish Europe by the advance it will make in France's favour'. A grateful Philippe was happy to give the go-ahead for Law's most ambitious idea of all. Law established a company, which came to be known as the Mississippi Company. In order to facilitate the system by which this company would be run Law required land. So Philippe obliged by handing over to Law France's largest colony, the Louisiana Territory, which occupied the entire Midwest of North America, from the Appalachians to the Rockies. The Mississippi Company would have a large issue of shares, and these would provide sufficient cash to mount an expedition to the French colony in America. The purpose of this expedition would be to mine the vast deposits of gold in Louisiana. No such opportunity had ever before been presented to the people of France. From the moment the Mississippi Company shares were first issued in June 1718 the country was gripped with gold fever. Within no time the shares were oversubscribed.

Up to this stage, Law had simply been putting into practice his brilliant, innovative ideas. Now his role becomes a little less clear. He couldn't possibly have had any evidence of gold deposits in Louisiana, because there weren't any. Yet all we know of John Law's character points against this being a massive fraud. It looks as if here once again Law was cannily playing the odds. South America had already proved a veritable gold mine for the Spanish – and a silver mine and source of semi-precious stones to boot. Who was to say that the vast unexplored tracts at the heart of North America wouldn't prove to be the same? Indeed, the rash bet would have been against such an eventuality.

The simple fact is that no one knew what Louisiana held in store. The territory had only been discovered and claimed for France less than forty years previously in 1682. Those who knew about these things in Paris were generally agreed that Louisiana was a large island off the coast of America, which either contained, or was near to, the mouth of a river called the Mississippi.

The financial district of Paris centred on the narrow, winding streets around the Rue Quincampoix. This attracted a colourful mix of loan sharks, government bond dealers and a variety of characters providing minor banking services. Here, they set up tables in the street, operated in the little restaurants and cafés, or in the warren of tiny rooms above. The Rue Quincampoix now became a bedlam, as people from all over Paris fought for the opportunity to buy shares in the Mississippi Company. As the price of the shares soared, demand became even greater. Shares began changing hands at a dizzying rate, and large amounts of money were soon being made by an increasing number of people. Many welcomed this new development. According to the contemporary diarist Jean Buvat, 'You see doctors of the Sorbonne, priests and nuns who mix in this business with everyone else. The world is enchanted by it all.' Here was a levelling process to challenge the rigid social hierarchy which had stifled France for so long. Women even began selling themselves for the opportunity to buy shares. Aristocrats had their servants set up chairs at the entrance to the Rue Quincampoix, so that they could observe the fun. Soon people were flooding in with their savings from all over France. As word spread, investors even began arrriving from Amsterdam and London.

Yet not everyone was quite so impressed by Law's new ideas. One of the few dissenting factions was the Parlement, which did its impotent best to oppose the monarchy. It was this body which had provoked Louis XIV to his famous remark: '*L'État c'est moi*' ('I am the state'). But in financial matters the Parlement remained hidebound conservative. It also reflected the xenophobic feelings of the French people. When its members learnt that they were to be paid in banknotes, rather than the customary gold, this was the last straw. A furious debate ensued, in the course of which it was decided that a party of armed guards should be despatched to seize the impudent Scotsman and bring him before the Parlement. Here he would be tried, sentenced to death and executed without further ado. Fortunately, Law got wind of his impending 'trial', and swiftly took refuge with Philippe in the Palais Royale until things blew over.

The Banque Royale, and its linked Mississippi Company, prospered beyond even Law's greatest expectations. He had promised Philippe that initially the Mississippi Company would generate

500 million *livres* profit; in fact, it produced well over three times that figure at the outset. With his monopoly bank and his monopoly trading company, Law found himself in possession of a massive and multiplying fortune. The Banque Royale opened offices in half a dozen provincial cities from Amiens to Lyon. In recognition of his services, Philippe created Law 'le Duc d'Arkansas'.

The first American duke moved into a sumptuous mansion on the Place Vendôme. This quickly became the focus of Parisian social life, with Lady Catherine becoming the leading society hostess. This is where the papal nuncio was so pleased to be invited to the seventh birthday party of Law's daughter that he gave her a kiss; meanwhile her teenage brother John went hunting with young Louis XV.

Law now had no time for gambling at the tables: his gambling was done elsewhere, and for much higher stakes. (As he put it: 'Commerce is a gamble on the future.') However, on the few occasions when he had time to himself he was pleased to keep up with one of his former hobbies. As his old conquest Elisabeth Charlotte d'Orléans tartly observed: 'Today M'sieur Law's hand was kissed by a duchess. There's no telling what part of him other ladies are liable to choose.' (The gossipy letters of Elisabeth Charlotte, with their blend of innuendo and grossness, written in a peculiar mixture of German and French, catch exactly the racy flavour of these times.) The portraits of Law dating from this time depict an elegant, bewigged figure in a superb braided topcoat, fashionable knee-length breeches and silk stockings. But even the portraitists' flattery cannot disguise his increasingly podgy features. High living, and the stress of his novel responsibilities, were beginning to take their toll on his handsome features.

It was during this period that Law began amassing his dozen or so châteaux and their estates, buying up tracts of land ripe for development at the edge of Paris, as well as entire city blocks and a large part of the fashionable Place Vendôme. Yet this was not simply personal extravagance. On the contrary, it was fulfilling two of his pet ideas – albeit in somewhat contradictory form. Here was money as action; it also gave the Banque Royale solid assets, turning it partly into a land bank.

This was not the only contradictory element which now began to emerge in Law's behaviour. As the figures expanded and things began running more and more towards the limits of control, Law's nerve became tested to the limit. His normal detached behaviour, the calculation of the gambler and seducer, now increasingly gave way to displays of unseemly emotion, as he found himself plunging deeper and deeper into matters beyond his competence. By September 1719 over 380 million *livres* worth of banknotes had been printed and were in circulation, while the bank had less than a tenth of this sum in solid assets. What many now saw as displays of arrogance and megalomania were in reality manifestations of creeping inner panic. Such behaviour is all too understandable in someone who has no previous experience of running a permanent residence, let alone the finances of the largest country in Europe.

Meanwhile vast fortunes were being made on the Paris Stock Exchange, which had been transferred from the narrow streets around the Rue Quincampoix to the large open square of the Place Vendôme. It was now felt that the all but unregulated exchange was in need of some order. To this end, trading was brought to a halt each evening by armed soldiers, who cleared the open square, and then mounted guard at the entering streets until the following morning – when they simply withdrew and the bedlam was permitted to restart. (Behaviour on the floor of modern exchanges has a long and glorious traditional precedent.)

Precisely how much money was being made? Historical monetary equivalents are notoriously difficult to pinpoint, but a rough estimate would make 500 *livres* equal to £250 or $400 at present-day prices. The figures involved were spectacular. By August 1719 the original 500 *livre* shares in the Mississippi Company were selling at almost 5,000 *livres*. The following month Law issued 100,000 new shares at a value of 500 *livres* per share. According to a contemporary source, the public 'fell on them like pigs'. The hysterical scenes can only be imagined. Anyone who bought just one share at the Mississippi Company's sumptuous new head-quarters in the Palais Mazarin, could immediately walk down the road to the Place Vendôme and sell it for 5,000 livres. People from all walks of life made hitherto unimaginable fortunes. A waiter made 30 million *livres*, a beggar made 70 million, a shopkeeper made 127 million. An aristocratic society hostess in the foyer of the Opéra found herself confronted by a woman whose *décolletée* was a dazzling cascade of diamonds. Raising her lorgnette, she was amazed to recognize her cook, and demanded what on earth she thought she was doing. The cook replied brazenly that she was now the equal of any aristocrat. A new word was even coined by the aristocracy to describe these people: they were referred to disdainfully as *millionnaires*. Members of the aristocracy too made money, though their attitude towards this commodity was charac-teristically different from that of the first millionaires. The Prince de Conti made nearly 5 million, some of which he spent on a diamond worth 5,000 *livres*. This he sent to a lady who had caught his fancy. When she returned his diamond he had it ground to dust, so that he could use this as writing sand to dry the ink of his letter replying to her.

Meanwhile Law continued with his bold financial strategy. This became known as the System, a word which was soon on everyone's lips. When the price of shares in the Mississippi Company finally broke the 5,000 *livres* barrier in September 1719, the System pulled off another coup. This effectively elimin-ated Law's last rivals, the four Pâris brothers. They ran their own rival Anti-System for generating money, which was known as the Tax Farm. This agency sold contracts for the 'farming' (collecting) of taxes in different regions. The more tax the contractor suc-ceeded in 'farming', the more profit he made. Indeed, the enthusi-astic methods used by the tax farmers made them particularly unpopular, and would in time prove a major factor in provoking the French Revolution.

However, with the advent of Law's System, people now preferred to invest their money here, rather than the Pâris brothers' Anti-System. As a result, the price of 'farming' sub-contracts tumbled, and in September 1719 Law managed to outbid the Pâris brothers for the Tax Farm contract. The Anti-System was incorporated into the System, and Law now had effective control of the entire government finances. The Mississippi Company had expanded its monopoly of French foreign trade from the Americas to the East Indies, and also ran the tobacco monopoly and the slave trade. Meanwhile the Banque Royale not only paid the government's debts, but collected its taxes. Besides printing paper money, it also had a nine-year contract to run the mint. In recognition of this *fait accompli* the Duc d'Orléans created Law 'Controller General of Finances', a post equivalent to prime minister. France was now being run by a foreign jailbird on the run for murder.

By this stage, however, the pressure of it all had driven Law to the very limit. Rumours spread of unbalanced behaviour behind closed doors at the Place Vendôme. Servants told of Law raging in his nightshirt through the empty salons, locking himself for hours on end in his closet. In public, displays of megalomania more frequently disturbed his normally detached demeanour. Any suggestion that his System might be wrong was angrily rejected. This was perhaps understandable. Law was indeed a man alone. No one else really understood what he was doing. Such ideas had never been tried before, and the attempts to control their effect were the first steps in an entirely new and untried science: one whose rules had yet to be discovered. Yet this was the only way. The truth or falsehood of such ideas – whether they were workable or misguided – could only be tested in practice. For this we owe a great debt to Law. His System is but the prehistoric forerunner of our own.

Law had good reason to be worried. Not everything was turning out quite as he had hoped, and ominous warning signs were beginning to appear. For a start, reports from the Louisiana Territory indicated that it showed little sign of living up to expectation. The coastal region was a dismal, alligator-infested swamp, whilst inland appeared to be mainly flat wilderness inhabited by fierce Indians. At present the territory had just a few hundred colonists, consisting largely of transported criminals: footpads, prostitutes, vagrants and the like.

Even the flat inland prairie hardly sounded like good mining country. However, vast tracts of the Territory still remained unexplored. What was needed was an expedition. Law had based his faith in the Mississippi Company upon the prosperity created in the Netherlands by the merchant traders. He still intended to make his company into another Dutch East India Company – though the spectre of the Darien Venture now began to haunt his thoughts. The trouble was, Law couldn't at present afford to mount an expedition to the Louisiana Territory. The income from the share issues had all been used to pay off government debts. And now the shares themselves were beginning to falter, as people started wondering about the promised gold mines. In an ingenious attempt to reassure wavering shareholders, Law secretly ordered the rounding up of all the beggars and *clochards* on the streets of Paris. These were then issued with picks and shovels, and marched through the centre of the city – continuing out along the road to La Rochelle, the disembarking port for the Americas. To all appearances, these were the men setting off for the Louisiana Territory to mine the gold for the Mississippi Company. The ruse appeared to work – though it wasn't long before the familiar bearded faces began to reappear in their old haunts, giving rise to suspicions in certain quarters. Yet for the time being shares in the Mississippi Company continued to rise – largely due, it seems, to their own self-generating momentum. (Five hundred *livres* shares would eventually peak at a stupendous 12,000 *livres*.)

There were also worrying signs at the Banque Royale. Some 2.7 million *livres* of banknotes had been issued by the bank, effectively doubling the money supply within a year. An astonishingly large proportion of this sum had been absorbed in stirring the sleeping giant of the French economy to life. The increase in spending on goods of all kinds – from market produce to furniture – had eventually generated full employment throughout the country. But now that this had been achieved, the rapidly increasing supply of money began chasing a very slowly increasing amount of goods. The excess money started to push up prices at an alarming rate. By the end of 1719 goods had risen by an average of nearly 75 per cent, with some essential foodstuffs in the Paris markets rising to three times their previous price.

Law had watched these developments with growing trepidation, but there was little he could do. Fortunately, other factors appeared to be balancing out such developments. In the previous July Law had taken steps to overcome any remnant doubts in the public mind about banknotes. At a stroke he had made coins less attractive, by the simple expedient of increasing the value of banknotes against coins. As the country's sole banker, Law could do whatever he wished with the currency. If he declared that a ten-*livre* note was worth double a ten-*livre* coin, then so it was. As has been pointed out by Law's biographer John Minton, this presented Law with a golden opportunity to set aside a small fortune for himself in Amsterdam or London – thus assuring that all his assets were not inextricably linked to the fate of the Banque Royale. But Law did no such thing. He was now the richest man in the world, yet he had achieved this only as a side-effect of proving his ideas. He regarded his financial achievement as very much secondary to his intellectual feat. All his assets were staked on his ideas. Law was the only person who knew this. Everyone else assumed that he was salting away large sums of money for himself in places far beyond the reach of France.

In February 1720 Law took a further step against coinage by

restricting all individual holdings of coins to 500 *livres*. But the novelty of banknotes was beginning to wear off, and the public's underlying suspicion concerning this form of currency began to increase. Although large fortunes were now being made in banknotes, their owners were quick to convert these paper fortunes into more solid assets. The price of jewellery, houses and land multiplied. Others began to invest in the booming markets of Amsterdam and London (where the share prices of the South Sea Bubble were now beginning to spiral). At every level it was felt that banknotes were for spending, not for saving: you saved gold; whoever heard of saving paper? As a result, the restaurants, theatres and bordellos of Paris prospered as never before. Consumption was all the rage.

All this seriously affected Law's calculations. The System relied upon a continuous flow of investment. At the same time his enemies were begining to conspire against him. The Pâris brothers, who were bent on revenge, remained a force to be reckoned with in financial circles. They were now joined by a powerful ally in the form of the Earl of Stair, the British ambassador. Stair was a former friend of Law, but he had taken against Law after sustaining heavy losses on the market. (This must be regarded as something of an aristocratic achievement, when commoners from valets to vagabonds were making their fortunes.) The Prince de Conti had also become irritated by Law, after Law's refusal to provide him with any more cut-price stock in the Mississippi Company. Conti decided that he had had enough: he was disgusted with the entire System. In a characteristic display of petulance, he despatched his entire holding of banknotes to the Banque Royale, demanding that it be replaced by gold. Guarded by armed soldiers, the three wagons containing Conti's gold made their way back through the streets to his residence. This hardly inspired public confidence in the new paper money, and might even have had disastrous consequences. Law recognized the potential danger and immediately sent word to Philippe at the Palais Royale. The Regent prevailed upon Conti, the gold was carted back through the streets to the Banque Royale, and a crisis was averted.

Law realized that the situation was in danger of getting out of control unless he acted decisively. To this end, on 21 May 1720 he issued a decree aimed at stemming the inflationary spiral of prices and at the same time reducing the excessive amount of paper money in circulation. This was a complex and ingenious measure, the like of which had never been tried before. It was also intended to be both gradual and just. Banknotes were to be devalued by 50 per cent, but this would not take place all at once – it would happen over a six-month period. Starting the next day, 22 May, banknotes would be devalued by just 20 per cent. Over the same period Mississippi Company shares, which now stood at 9,000 *livres*, would be reduced in price to 5,000 *livres*. This was an astute move. It wouldn't take long for people to realize that the devaluation of the shares was less than that of paper money. People would flock to invest their paper money in Mississippi Company shares, thus reducing the amount of banknotes on the market. Unfortunately, what Law didn't realize was that, by fixing the price of the shares, he was effectively turning these into money. Instead of going down, the money supply effectively went up.

The consequences of Law's decree were catastrophic. This was the first devaluation the public had experienced. (Few had realized that Law's previous 'revaluation' of the coinage was in fact a covert devaluation. Despite widespread involvement in the world of finance, most people remained naive about its actual mechanics.) The way the public saw it, this devaluation meant they were being robbed. Half their money was simply being taken away from them. In the words of the Duc de Saint-Simon, 'The uproar was general and frightful. Every rich man considered he was ruined beyond redemption; every poor man believed himself reduced to beggary.' In the ensuing panic everyone tried to get rid of both their banknotes and their shares in the Mississippi Company. The bubble had finally burst, and mayhem ensued.

In an effort to stem the deluge, Law decreed that to begin with only ten-*livre* notes would be redeemable. As a result, huge queues formed in front of the pay-out windows of the Banque Royale, their lines snaking down the nearby streets. Armed soldiers did their best to keep order, but this didn't prevent toughs from muscling their way to the front of the queue. Some are even said to have swung like monkeys through the branches of the trees lining the street to jump the queue. Fisticuffs and knife fights broke out, and several people, including one soldier, were killed in violent disturbances.

Soon the bank was being stormed by thousands demanding payment for their banknotes. According to a contemporary commentator, at one point there were 300,000 gathered in the streets around the bank. This is certainly an exaggeration, for it represents half the population of Paris at the time. Even so, the mob panic is difficult to exaggerate. The militia was mobilized to control the crowds. There were bayonet charges, and shots fired over the heads of the crowd to disperse the clamour. Despite this, on 17 July fifteen people were crushed to death in the press outside the bank. The Parlement now decided to step in, and revoked Law's decree limiting the holding of coins to 500 *livres*. Philippe replied by banishing the Parlement to Pontoise, in those days a small town beyond the northwestern outskirts of Paris. It was the end of the System. In the words of Voltaire, 'Paper money has now been restored to its intrinsic value.'

Law was described as 'trembling for his life', as well he might. After several days behind barricaded doors at the Place Vendôme, the front of his mansion protected by armed guards, Law and his family were permitted to take refuge in the Palais Royale. A few days later a woman mistaken for Catherine Knollys was thrown into a duckpond and all but drowned by the mob before she could be rescued. An army officer mistaken for Law himself was the subject of an assassination attempt. Feelings ran high through every section of the population. Just one thing was responsible for their plight: the System. And just one man was responsible for the System: John Law. Yet astonishingly, even now Law believed that his System could be made to work. He explained to Philippe that it had only failed through a freak combination of circumstances, in which public ignorance played a large factor. People just didn't understand what the System was, and what it was doing for them. It was the freak circumstances, rather than the System itself, which had brought about the currency crisis and the run on the bank. Even today, our 'System' remains vulnerable to much the same circumstances. During 1998, when Russian depositors lost confidence in the banks and queued outside to withdraw their deposits, the banks were forced to close their doors and many simply collapsed.

Despite his safe haven in the Palais Royale, Law was afraid he might be assassinated, or even executed. Yet Philippe remained loyal to the man who had tried to do so much for him and his country. And in certain ways Law had arguably succeeded. Some historians even contend that the overall situation was now significantly better than before Law had taken over. The excess of paper money meant that almost everyone had managed to pay off their debts (many of which had been crippling and of long standing). The administrations of several provincial cities had lost fortunes, but the cities themselves had been stirred into a new commercial life. Also, the flight from paper money into solid goods meant that there were now many more property owners. At the same time the rigid social barriers of the *ancien régime* had been seriously undermined: the seeds of democracy had been sown. Many, like the cook with her diamonds, felt themselves the equal of their so-called superiors. In seventy years' time this feeling of *egalité* would combine with *liberté* and *fraternité* to become the French Revolution.

By September 1720 order had been restored in Paris. The enormity of what had happened had sunk in. France's financial system was bankrupted. An eerie calm settled over the city. People stood on street corners in dazed groups. So many had been forced to sell their carriages that there was hardly any traffic.

Three months later Law was encouraged to slip away to the Château Guermantes, one of several aquisitions he had made in the countryside outside Paris. During his glorious 500 days of power, his control over France had been more penetrating than that of the Sun King and its other absolute monarchs. Now Law was returned to his former existence. He was on the run again, in danger of his life.

Eventually Philippe granted a passport for Law and his sixteen-year-old son. But Catherine Knollys and their daughter were not permitted to leave the country. Law departed for Brussels, where he put out feelers about being allowed to return to Britain. He had now become a celebrity throughout Europe, and a pardon was quickly granted. (Doing a disservice to France is customarily regarded with favour in England.) Law's first appearance at a London theatre created a sensation: the Prince and Princess of Wales in the royal box were ignored as all craned to stare at 'the Scotchman who has bankrupted France'.

Law had salted away no money of his own. All his property in France had been seized, and he owed millions. Acting in France's interest, Law had dealt privately with bankers in London, Amsterdam and Hamburg. In this way he had obtained personal loans, which had been used to service and oil the wheels of the System. The accumulated total of these foreign debts alone was 6.7 million *livres*. From being the richest man in the world, Law had now plunged headlong into the abyss of the financial realm. Yet even in his extremity he continued to petition Philippe. The System was France's only hope: his mind was brimming with new monetary ideas. In his view, the System had not yet been fully implemented. There was more to come . . . But Philippe was in no mood to listen. His years of excess were catching up with him. In three years he would be dead.

Law too was but a shadow of his former self. Destouches, the

French ambassador in London, described him: 'He appears a dreamer, pensive, broken and distracted.' Gone was the charmer of princes (and their women). The man of cool calculation and dazzling innovative ideas was reduced to an ageing trembler with a pronounced tic. In London he turned fifty years old.

In 1724 Law went to live in Venice. Here, he supported himself by gambling at the Ridotto casino. Cannily he eked out his last assets, cashing in on his fame and his knowledge of the odds. There were many who wished to boast that they had gambled with the notorious Mr Law. He would sit at his table, by his elbow ten thousand *pistoles* of stake money laid out in little stacks of coins (no notes!). Law would challenge any comer to throw six dice and come up with six sixes. Against his challenger's outlay of one single gold *pistole*, he would wager his entire stake. Odds of 10,000–1. One by one he collected his single gold *pistoles* from those who were tempted by the odds. (Law knew that the real odds were 46,656–1.) His fame made him an object of curiosity. The passers-by would stare at the brooding figure, in black tricorn hat and black cloak, hunched over a café table in the Piazza San Marco. He was not well, and the mists which rose from the canals and seeped throught the alleyways were not good for his lungs. Yet still he would be sought out by passing celebrities. The 35-year-old *philosophe* Montesquieu had satirized Law viciously in his *Persian Letters*. Holding Law up to ridicule, he had thinly disguised him as the son of Aeolius, the god of wind, travelling the world with the blind god of chance and an inflated bladder. Yet on Montesquieu's visit to Venice he was both surprised and impressed to find Law, 'his mind occupied with projects, his head filled with calculations'. But this must have been on a good day. Others spoke of Law's spiritual and bodily exhaustion, his formerly handsome face now haggard – with oafish, slack lips and an increasingly distressing tic. One night at the end of February 1729, returning home on a gondola through the chill mists of the canals, he caught a 'distemper'. This soon developed into pneumonia, and within a month he was dead.

So was Law a fraudster? The Earl of Stair, who knew him well during the period of their friendship, said Law 'had a head fit for calculations of all kinds to an extent beyond anybody' and reckoned that he was 'the cleverest man that is'. Some claim, in the age of Newton – of whose intellectual pre-eminence even Stair must have been aware. Just over a century later Marx would refer to Law as 'the pleasant character mixture of swindler and prophet', whilst in the twentieth century the great Austro-American analyst of capitalism Schumpeter would place Law 'in the front rank of monetary theorists of all time'. In essence, Law's System consisted of a few brilliant ideas: money as action, fiduciary issue of banknotes, the benefits of merchant trading and so forth. These were supported by an understanding of the mechanics of finance well ahead of its time. What unfortunately Law did not understand – and how could he have understood this? – was the instability of such a system in its pioneer state, when first put into practice. The 'system' by which the financial world runs today is based on similar foundations, but it is hedged about with a host of checks and balances designed to alleviate this instability. They do not eliminate it. Instability would appear to be fundamental to any such system, and is arguably the very nature of how it works.

3

Before Adam

1720 was to prove a date to remember throughout the European financial world. As well as Paris, the two main exchanges in London and Amsterdam also experienced their own traumatic events. What happened in London is now known as the 'South Sea Bubble'. Some economic thinkers consider this event to be of little serious interest. Adam Smith, no less, was to dismiss it as a 'meer fraud'. Yet there is no denying that it has its place in the allegory of monetary progress (lying somewhere on the path between the Rainbow of Optimism and the Vale of Ruin).

The South Sea Company had been founded in 1711 by a mixed group of financiers with strong connections in parliament and the City (London's financial heart). The driving force behind this group was an ambitious legal scrivener called John Blunt. By all accounts a burly, domineering character of ingenuity and ambiguous charm, Blunt was the son of a shoemaker who had taken a large step up the social ladder by marrying into a respectable county family. This was a time of social change. The newly united kingdom of Great Britain was emerging as a European power, intellectually, militarily and commercially. Newton was acknowledged as the supreme genius of the age, while London's Royal Society was seen as its supreme scientific arbiter. Locke was founding empirical philosophy and promulgating the liberal political ideas which would, by the end of the century, be embodied in the American constitution. Meanwhile Defoe's *Robinson Crusoe* and Swift's *Gulliver's Travels* would cater in their own separate ways to the public thirst for foreign adventures. This was a self-confident nation, experiencing the earliest stirrings of what would become the Industrial Revolution – already steam power was being used in the mines of Cornwall. Amidst the rising tide of commercial activity new fortunes were being founded. John Blunt and his associates were determined to ride the crest of this wave, and saw the South Sea Company as their big chance.

The South Sea Company was a gamble from the start, though it must be admitted that commentators with hindsight vary in degree on this opinion. Those who side with Adam Smith maintain it was always a sham; others, also expert and in possession of the same facts, maintain it started as a rash idea and was simply overtaken by circumstances. However, like John Law's System, it was undeniably a rite of passage for the financial world, which would emerge chastened and in some ways matured. If it hadn't happened at this particular time in this particular manner, something very similar would certainly have occurred, as indeed it has done at regular intervals since. This rite of passage seems to be a repeating test of survival, and should not be mistaken for a learning process.

The South Sea Company was founded as a trading company. In exchange for taking on a hefty portion of the national debt (£9.5 million) in the form of unsecured government loan stock, the government granted the company monopoly trading rights to Spanish South America 'on the east side from the river Aranoca to the southernmost part of Tierra del Fuego' as well as the entire west coast and anything there 'which shall hereafter be discovered'. The government also guaranteed to pay the South Sea Company an annual rate of around 10 per cent (£600,000) on the debt stock. (To give an idea of the value of this payment, at the time a middle-class family could live well on £200 a year, anyone with twice this income was considered rich, and a manservant cost £10 a year in wages.)

Unlimited access to the gold and silver mines of Mexico and the Andes, along with a guaranteed income of over half a million

– as soon as the company had an issue of stock, there was no shortage of takers. There were, however, a few snags – which should have been plain to all. The War of the Spanish Succession, which had begun in 1702, was still dragging on. The Spanish had not yet lost, which meant they had not yet been forced to sign any treaty giving away their monopoly trading rights to their South American empire. And in the opinion of Defoe, this would not happen 'unless the Spaniards are to be divested of common sense'. Another drawback was the fact that none of the directors had any experience whatsoever of trade with South America, or even any notion of South American affairs. (One supects their geographical idea of this spot was probably similar to contemporary Parisian notions concerning the island of Mississippi.) The sole exception to this ignorance was Blunt, who had a cousin in Buenos Aires. Yet this hardly augured well. Blunt's main contact with his cousin had been to lend him £100, which had never been repaid.

In 1713 the War of the Spanish Succession duly came to an end. The Spanish were forced to sign the Treaty of Utrecht, in which they gave up their monopoly of trade with their South American empire. The South Sea Company would be permitted to send one ship per year for general trading, on condition the Spanish were granted a share of the profits. This news does not appear to have dispirited stockholders in the South Sea Company. Even the fact that the government had fallen behind with their promised annual payments didn't dent their confidence (the company remained in the red until 1716). Finally, after much wrangling with the Spanish authorities, in 1717 the first South Sea Company ship returned from its trading voyage to South America. The resultant modest profit served only to boost the general confidence, which was further inspired a year later when no less a figure than George I accepted the governorship of the South Sea Company.

The South Sea Company was allowed to issue stock up to the value of the national debt which it had taken on. Blunt and his directors decided it was time to issue more stock. But to do this they needed to take on more of the national debt, which required an act of parliament. Influential cabinet ministers were quietly promised allocations of new stock to ease the passage of the bill; and the members of parliament who were directors or simply held stock made impassioned speeches pointing out the benefits of this move. As a result, in 1720 the South Sea Company was allowed to take over the entire national debt (some £51 million). The ensuing boom in South Sea stock exceeded even the most optimistic expectations. In January 1720 £100 stocks had been selling for £129. Three months later this same stock was selling for £330; by May it had reached £550. Chaotic scenes similar to those in the Rue Quincampoix now began taking place in the narrow streets and lanes of the City.

This was the year that the 23-year-old Hogarth finished his apprenticeship and struck out on his own as an artist. One of his early engravings is called *The South Sea Bubble* and depicts a chilling scene. Investors spin on a whirligig above the heads of the gaping crowd, some of whom wait by a ladder for their turn to climb up for a ride. At the edges of the crowd the naked female figure of Honour is being broken on the wheel by Self-Interest; Honesty is being flogged with a cat-o'-nine-tails by a beturbaned Villainy; and the Devil is slicing slabs of flesh from the hanging naked body of Fortune, tossing them to the brawling crowd below.

Fortunes were made by many speculators. Butchers, lawyers and members of parliament all reaped the benefits of the sky-rocketing price of South Sea stock. Even the English landed aristocracy, to whom all 'trade' was vulgar, were soon swallowing their pride and investing. In June 1720 South Sea stock rose to a staggering £890. By now financial operators in the City had realized they were on to a good thing, and hundreds of similar joint stock companies had sprung up. Some of these were for purposes which with hindsight appear far more plausible than those of the South Sea Company, such as dealing in hair for wigs, the insuring of sheep herds and the like. Others were admittedly less plausible:

one company was set up for the manufacture of a perpetual motion machine, and another 'for carrying on an undertaking of great advantage, but nobody to know what it is'. (Those who find this amusing presumably do not own shares in Monsanto, Rio Tinto, or any of the pharmaceutical multinationals who spend millions on research into 'nobody to know what it is'.)

People carried on investing regardless. Others were not so happy at this development. The so-called Bubble Act was rushed through parliament to prevent the setting-up of such companies and to protect the innocent speculator – at the same time protecting the near monopoly which the South Sea Company had on such speculators' funds.

Meanwhile the boom in the City of London sparked similar booms in the stock markets of Amsterdam and Hamburg, where companies could still be set up without interference from the Bubble Act. In August South Sea stock rose to a giddying £1,050. But this was to be the pinnacle. A month later the bubble burst. Contemporary commentators disagreed as to which pin actually pricked the bubble. Some blamed the collapse of so many similar, but utterly fraudulent companies as a result of the Bubble Act. Others suspected covert large-scale profit-taking by directors and their parliamentary cronies. Yet others suggested the laws of gravity were responsible. It was probably a combination of all these, and more.

By September South Sea stock had plummetted to just £175. Amidst the bedlam of panic-selling, valiant efforts were made to rally confidence. Passionate and patriotic speeches were made in parliament, doubtless by those members who still retained stock. Ministers made official pronouncements in the manner of Canute; voices in high places vainly called for the South Sea Company to be rescued by the Bank of England. (Had this come about, the recently founded Bank of England would undoubtedly have gone the way of Law's Banque Royale.)

The sheer volume of voices expressing confidence had but a

momentary effect. For a while South Sea stock bottomed out at around £140. Directors of the company pointed out that this meant anyone who had bought stock the previous January at £129, and hung on to this throughout the bubble, had still made a profit of £11. Such consolations fell on deaf ears. The market rarely works this way. Everyone had bought and sold. Massive expectations had been roused. Assets had been hocked and mortgaged, money had been borrowed and promissory notes written – all in order to buy stock which in some cases had lost more than 80 per cent of its value. There was widespread devastation and ruin amongst the investing classes – which had now expanded considerably. Tradesmen and landed gentry, professional men and members of parliament – all suffered severe losses. Even Newton lost out to the tune of £20,000. The tidal wave of selling soon hit the exchanges in Amsterdam and Hamburg. 1720 had proved an *annus horribilis* throughout the European financial world.

Who was to blame? In London a parliamentary inquiry was set up, and four ministers were found guilty of accepting bribes, as well as using privileged knowledge to further their own gain. Noisy public meetings throughout the country demanded prosecution for the South Sea Company directors. The hunt for scapegoats was in full cry. Sir John Blunt (gratefully knighted during the expansion) was the victim of an assassination attempt: an angry debtor tried to shoot him in the street. Blunt at once took refuge, and began naming names in high places who had connived. Members of parliament who had dealt in shares were expelled, and the government fell. Directors of the company had their assets and estates seized to provide compensation. A senior statesman committed suicide; and the company treasurer, Robert Knight, caught a boat for the Continent, hotly pursued by a warrant for his arrest. Everyone blamed everyone else. No one would accept the most obvious fact: everyone was to blame. Call it greed, call it collective hysteria, call it gullibility . . . None had been immune: not even Newton.

The twentieth-century economist J. K. Galbraith cites the South Sea Bubble as an exemplary case of its kind, embodying all the elements found in any such episode. Amongst these he lists the following: 'Individuals were dangerously captured by belief in their own financial acumen and intelligence.' This caused them to pass on this contagious delusion to others. Another 'predictable feature' was 'an investment opportunity rich in imagined prospects' which simply didn't stand up to realistic assessment. And finally: 'the mass escape from sanity by people in pursuit of profit'. Self-convinced astuteness mounting to collective hysteria: the ingredients have proved the same from 1720 to 1929, through 1987 and beyond to the first dot.com collapse of the twenty-first century.

If ever there was a need for a hard-headed and instructive parable on social behaviour, this was the time. Such was to be provided by Bernard Mandeville in *The Fable of the Bees*, which became a best-seller in the aftermath of the South Sea Bubble.

Mandeville was born in Holland in 1670, and had taken a degree at the medical school in Leiden, then the best in Europe. He came to England in his mid twenties to learn the language, but he 'found the Country and the Manners of it agreeable', and stayed on, soon becoming married to an Englishwoman. His medical practice does not appear to have flourished, and at one point he undertook public relations work for a distillery. Gin had been invented in Holland just a few decades previously, and this cheap spirit was now being introduced with disastrous effects into England. The gin palaces where one could get 'drunk for a penny, dead drunk for tuppence' would later inspire the squalid vision of Hogarth's *Gin Lane*.

Mandeville's medical specialization was 'Hypochondriack and Hysterick Diseases', which may account for his lack of prosperity in an age where diseases still tended to be all too real. (The Plague was still within living memory, and the Bills of Mortality differed little from Graunt's day.) However, Mandeville's unusual specialization does imply a certain psychological acumen, and this he was to use to the full in his masterwork, whose full title was *The Fable of the Bees, or Private Vices, Public Benefits*. This was published in pamphlet form, and consisted of a long doggerel poem, together with various 'Remarks' expounding its ideas. These were intended as nothing less than 'An Enquiry into the Origin of Moral Virtues'. As such, they succeeded in explaining how we work as social beings better than any philosopher since Machiavelli, two centuries previously.

Mandeville's poem uses a hive of bees as an image of human society. He then describes how this flourishing hive is reduced to poverty and depopulation by an outbreak of virtue. When the bees begin behaving in a selfless manner, indulging in self-denial and curbing their individual self-interest, their society begins to fall apart. When the individual bees no longer seek to accumulate a wealth of honey this spells ruin for the hive.

Mandeville's cardinal insight was that social progress and prosperity have nothing to do with individual virtue. On the contrary, such vices as greed, ambition and vanity are what drive a society to prosperity. Self-interest, not selflessness, is what brings social benefit.

> What we call Evil in this World, Moral as well as Natural, is the grand Principle that makes us social Creatures, the solid Basis, the Life and Support of all Trades and Employments without Exception: That there we must look for the true Origin of all Arts and Sciences.

Indeed, 'private Vices are public Benefits'.

This somewhat pessimistic view of human affairs was a necessary corrective for the times. It deepened social self-understanding. Prior to this a virtuous citizen may well have experienced a moral queasiness with regard to his commercial actions, detecting a certain hypocrisy in his practice of them. Yet he would not fully have understood why he felt this way. Mandeville articulated this situation, with all the effrontery of Machiavelli's advice on how to succeed in politics. Even so, Mandeville's nihilistic view is not the whole story. His view of virtue was decidedly narrow, very much a legacy of puritanism. Virtue, in the personal sense, is more than simple self-denial. Likewise, Mandeville's conception of personal vice was similarly narrow. Not all the colourful or forbidding characters who ended up swinging from a rope at Tyburn were social benefactors in disguise.

It comes as little surprise that Mandeville's moral thesis caused widespread outrage amongst upright citizens, especially those in business. In 1723 *The Fable of the Bees* was even censured as a 'public nuisance' by the grand jury for the county of Middlesex (which at the time included the whole of London). It is only with hindsight that we can recognize the further extent of Mandeville's originality. His insight into the mechanics of the market was profound, if not systematic. In his view there was no need for any interference in these mechanics, either by government regulation or individual altruism. The market was best left to its own devices. 'As things are managed with us, it would be preposterous to have as many Brewers as there are bakers, or as many Woolen-drapers as there are Shoemakers. The Proportion as to Numbers in every Trade finds itself, and is never better kept than when nobody meddles or interferes with it.' Here was the first reasoned advocacy of economic laisser-faire. Mandeville also understood the important point that the 'division of labour' made for increased efficiency. Indeed, it was he who first coined this term. Mandeville understood the benefits of laisser-faire, but he also saw its cruel side-effects. 'To make Society Happy ... it is requisite that great numbers should be Ignorant as well as Poor.' This has been cited as Mandeville's 'Machiavellian strain' at its worst. Such may well be the case, but it also introduces another key element into economic thinking. What to do about the poor? Mandeville's understanding of private vices, and his introduction of the problem of the poor, bring what is for many non-economists the most fundamental issue of all into the economic arena. Namely, the question of economic morality. Mandeville can hardly be mistaken for a moral thinker: his amorality was undeniably Machiavellian. But his insight into morality (private vices) and his rejection of it (with regard to the poor) meant that from now on economics and morality were inextricably linked. Economic thinking was seen to involve the interests of all members of society, not just 'the state', the rulers, the merchants or some other vested interest. For many years no one would know quite how to accommodate morality and economics, especially with regard to the poor. How was it possible to make an economic system just for all? Ensuing economic thinkers may have ignored this question, or simply despaired of it – but the question had been raised. What Mandeville had joined, no man could cast asunder.

From morality to mathematics: a revolutionary discovery was now made at the other end of the economic spectrum. For the first time this linked the random disorder of the world (of objects as well as people) to a numerical principle. The discovery was made by a crochety mathematician called De Moivre, who lived and worked in London at the same time as Mandeville, and almost certainly met him.

Abraham De Moivre was born a Huguenot (Protestant) in France in 1667. De Moivre had shown exceptional early promise at mathematics, but this was scotched in 1685 when Louis XIV revoked the Edict of Nantes, which had guaranteed Protestants equal political rights with the majority Catholics. De Moivre was imprisoned for three years, and then fled to England. Despite his brilliance, he was unable to obtain an academic post, and was forced to eke out an embittered living as a mathematics tutor. For the most part his pupils appear to have been rich young recalcitrants or enthusiastic ignoramuses of more mature disposition. De

Moivre discovered Newton's *Principia*, and is said to have torn individual pages from his copy to memorize while walking the streets between his tutoring appointments. When his work was finished for the day, he would retire to his regular table at Slaughter's Coffee House. Here he earned a few more pence passing on advice concerning probability to gamblers, insurance brokers and other freelance financial operators of the period. The latter were particularly in need of such advice, and not just concerning the investment probabilities of the South Sea Company and the like. The relation between probability and actuality still remained hazy in the extreme. An instance will suffice. In the early eighteenth century it was possible to purchase a life annuity from the British government. Nothing extraordinary in that – except that the size of the deposit and the size of the annual payment were the same for all. Age was not regarded as a pertinent factor in this arrangement.

As ever, canny freelance financiers were one step ahead of the government, and were only too pleased to consult with M'sieur De Moivre at Slaughter's Coffee House. This was the great era of London coffee houses, where a wide strata of society met, exchanged gossip and contracted business. In the absence of newspapers as we know them today, this was where one heard the latest news concerning wars, business ventures or even shipping movements. Samuel Pepys, when he was Secretary to the Admiralty, mentions in his *Diary* how he would call in at Lloyd's Coffee House on Lombard Street to learn which of his vessels had arrived in the Port of London. (When customers took to brokering insurance contracts on ships in this coffee house, it was the beginning of the worldwide insurance exchange now known as Lloyd's of London.) Commercial activity of all kinds was booming, as well as breaking into innovative fields such as insurance. Clients sought out insurance brokers in the coffee houses, where they would draw up contracts against such things as robbery by footpads, death from a surfeit of gin, 'collapse of abode', or even 'assurance of female chastity'. For the cost of an agreed premium the value of the loss would be covered. The broker would then tour the coffee houses and sell on the risk to any individual willing to write his name under the agreed contract. In this way individual insurers acquired the name of 'underwriter'. Such were the financial risk-takers who consulted De Moivre on probability.

Despite his misanthropic ways, De Moivre's regular visits to Slaughter's Coffee House meant that he encountered several of the leading figures of his day, and at the same time achieved no little renown for his mathematical skills. In his epic poem *Essay on Man* Alexander Pope, the poet *par excellence* of this age, refers to a spider's innate geometric abilities in spinning his web as being 'sure as De-moivre'. Both Dr Johnson and the astronomer Halley (after whom the comet had just been named) were impressed by De Moivre. The latter would even go so far as to introduce him to Newton, who was not easily impressed. Yet, when questioned by a colleague about probability, the grand old man of English science would reply, 'Go to Mr De Moivre; he knows these things better than I do.' On the recommendation of such peers De Moivre would be elected a Fellow of the Royal Society at just thirty years of age, and almost forty years later he would also be elected a fellow of the Berlin Academy of Sciences. Yet he never rose above being an impoverished hack tutor and coffee house advice-pedlar, which suggests more than some ordinary difficulty of character. Cambridge wouldn't have him (despite a recommendation from the great German philosopher Leibniz); Oxford was persistently not interested; and Leiden didn't even ask. De Moivre was to remain a crabbed local character on the coffee house scene for over sixty years.

The decades of meagre living eventually took their toll. By the time he had reached seventy, he was described as 'almost fit for his coffin; he was a mere skeleton, nothing but skin and bone'. But he kept going well into his eighties. In his final years the aged, peripatetic tutor was said to have become overwhelmed with lethargy (as well he might). He would lie in his bed sleeping ten, twelve, fourteen hours a day. According to a joke which went the rounds, he would sleep a quarter of an hour longer each day, and would be dead when he finally slept the whole day through. Abraham De Moivre finally died at the age of eighty-seven, having made a contribution to human understanding which would never be forgotten.

De Moivre's contribution lay in the question of probability, which had first been raised over two centuries previously by Pacioli's unfinished game of balla. Others such as Pascal and Fermat had built upon this, but it would be De Moivre who made the major breakthrough which would later play such a role in economic forecasting. De Moivre dealt with the totally random problem of tossing a coin. Each time the coin was tossed, there was no way of telling whether it would land heads or tails. And each toss of the coin was unaffected by any previous toss. It was always a matter of utterly independent chance which way the coin landed: the random element remained unaffected, no matter how many times the coin was tossed. If, for example, the coin was tossed 100 times, it was likely to end up 50 heads and 50 tails – but this was far from being certain. Sometimes it might be 48:52, at other times it might even be 41:59. But more times it would be closer to the average 50:50. De Moivre puzzled over this probability, but it was not until 1733 that he finally understood how this series of utterly independent random events followed a regular distribution pattern. The deviations from the average 50:50 followed a distinct pattern in their magnitude and frequency. This pattern came to be called 'normal distribution'. It can be represented on a graph, where it manifests itself as a 'bell curve'.

The curve is shaped like a bell, and is symmetrical either side of the mean, which is 50 heads (and, of course, 50 tails). In each series of 100 tossings the most frequent occurrence will be the mean, which is the highest point of the curve. The next most frequent will be the closest to the mean, such as 49 heads or 51 heads. The frequency

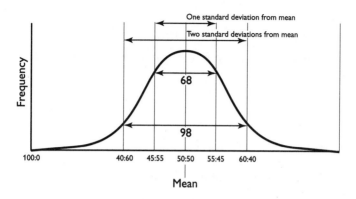

*Bell curve for occurrence of heads and tails each time a coin
is tossed 100 times*

declines sharply as the number of heads deviates further from the
mean (the curve sloping steeply downwards). But when we
approach the furthest deviations from the mean, the curve levels off
at a very low frequency (like the rim of a bell). In other words, you're
almost as likely – or unlikely – to get 80 heads as you are 90 heads.
Both of these are hardly likely at all!

De Moivre discovered a highly revealing characteristic of
normal distribution, which he called 'standard deviation'. He
found that in normal distribution a fixed number of occurrences
will always fall within a definite limit either side of the mean.
For example in a series of 100 coin tossings, 68 per cent of
the time the number of heads will fall within one standard
deviation from the mean. This accounts for the bulge of the
bell. He also discovered that 95 per cent of the time the
number of heads will fall within two standard deviations. But what
exactly are these 'standard deviations'? De Moivre worked out
that the standard deviation for a coin which is tossed a series of n
times is:

$$\frac{\sqrt{n}}{2}$$

For example, when a coin is tossed 100 times, the standard deviation is:

$$\frac{\sqrt{100}}{2} = \frac{10}{2} = 5$$

Therefore, the standard deviation either side of 50 is 5. So for a series of 100 – each one of the series consisting of a coin tossed 100 times – 68 per cent of the number of heads will lie between 45 and 55. Two standard deviations are $5 \times 2 = 10$. So when the coin is tossed in a series lasting 100 times, 95 per cent of the number of heads will lie between 40 and 60.

Here was a pattern which lay at the very heart of randomness: an order in the midst of disorder. In De Moivre's eyes such a marvel – such an impossible contradiction – could only be God's work. It must have been part of the Creator's initial conception of the universe: 'that Order which naturally results from ORIGINAL DESIGN'.

The bell curve is now seen as the identifying mark of randomness. It occurs wherever randomness is encountered – from the tossing of coins to the distribution of IQ amongst professional wrestlers (or mathematics professors). Indeed, it physically manifests itself in the wearing-out of ancient stone steps. Even when events were random, it was possible to predict the probability of what would take place. Here was just the tool that was needed for 'political arithmetick'.

Theoretical insights – such as those provided by Mandeville and De Moivre – developed largely in Britain. Here, the eighteenth century was an era of widespread scientific advance and commercial enterprise presaging the Industrial Revolution. Yet curiously, the first systematic economic thinkers appeared in commercially backward France – where in the aftermath of the Law débâcle people remained understandably suspicious of such modern aids as paper money or banks. This group of innovative French thinkers came to the fore in the latter half of the eighteenth century and were known as the Physiocrats. Indicatively, they often referred to themselves as '*economistes*', and were the first to think of this word as we do today. (The word 'economics' in fact originates from the ancient Greek *oikonomos*, meaning 'the running of the household'.)

France at this time remained largely agricultural, under the centralized autocratic rule of Louis XV at Versailles. (John Law's pal, the roué Regent Philippe II, had lasted just long enough to hand over a ruined nation when Louis XV came of age in February 1723.) France now existed in a state of uneasy stasis. The intellectuals of the Enlightenment, such as Voltaire and Rousseau, promulgated their liberal ideas but were unable to act. Many were beginning to think long and hard about the structure of France and how it worked. Yet it was France's very stasis, its agricultural simplicity, which made it ideal for the first systematic economic analysts.

The leading Physiocrat thinker was François Quesnay, who was born in Paris in 1694. His family had long been poor, but François' father made sufficient as a small merchant to bring up his family in some comfort. Despite this, young François received no early education and remained illiterate until he was ten. He then developed a passion for medicine, which appears to have combined with an intuitive talent for assessing a patient's needs. His exceptional medical skills were underpinned by a firm belief in nature's power to heal the body's ills. Medical practice was still very much in the sawbones, leeches and elixirs era. As a result, Quesnay's hands-off approach quickly brought him extraordinary success – presumably on the principle of less done, less harm. In 1749 he was appointed personal physician to Madame de Pompadour, the powerful and talented mistress of Louis XV. Quesnay took up residence at Versailles, and three years later he succeeded in curing the Dauphin (the heir to the throne) of a potentially lethal dose of smallpox. Quesnay was rewarded with a title and a country estate. Running this estate appears to have stimulated Quesnay to thoughts about running the state as a whole. At the late age of sixty he began applying the systematic scientific understanding he had gained in medicine to the larger estate of France. The result was a body of thought whose key concept remains central to this day.

Physiocratic ideas are a curious mixture of the progressive and the archaic. At their centre is the notion of economic and individual freedom, yet this is envisaged in a country with a rigid, unchanging structure. Quesnay's thought contains three basic concepts. Firstly, the social and economic scheme of things was governed by natural law (*le droit naturel*). This was preordained, and in many ways identical to Divine Providence. Yet it was at the same time a subtle and complex social mechanism which was inherent in the very working of things. The protection of life and property were in accord with the natural law, as was the freedom to buy and sell. If trade was permitted to continue freely in this way, the natural law would run its course. Things would thus be able to work out in their own fashion, and develop in accordance with what was best. Even amidst such a hazy conception, the intrinsic circularity shines through: what happens in the market is best for the market; what is best for the market is what happens in the market.

Yet this circularity was of little concern to Quesnay. The stasis in France led him to assume a state of economic equilibrium, which meant the market was not likely to change in any drastic fashion. (This assumption was to have dire consequences for ensuing economic theorists, who unthinkingly made the same assumption while attempting to legislate for economies that were developing.)

So the best thing for the market, according to Quesnay, was to leave it alone. This was embodied in his second basic concept:

laisser-faire (let it be). Here was a concept which would play, and continues to play, a central role in economic thought. Quesnay's laisser-faire, which was closely allied to his idea of natural law, had distinct theological implications. Modern versions of laisser-faire, besides being less absolute, tend to see the competitive workings of the market as merely better than government regulation. The market itself is the ultimate guiding force, not God (though the astute will have spotted that this eludes neither circularity nor a belief in some quasi-ineffable providence).

As we have already seen, Mandeville had previously advocated something very similar to Quesnay's laisser-faire. However, Quesnay was the first to name it as such, and the first to introduce it as a central notion within a body of systematic thought. Quesnay's advocacy of laisser-faire was in sharp contrast to the loose body of ideas known as Mercantilism, which remained the prevailing orthodoxy. The Mercantilists had believed in regulating the market so as to foster the interests of national merchants. Protective measures were advocated against foreign imports. The market was regulated in favour of merchant guilds, which maintained professional standards. The Physiocrats, on the other hand, argued that such restraint was against the natural law which underlay all trade. It also fostered monopolistic tendencies, and was even short-sighted where the interests of the merchants were concerned. Laisser-faire would enable efficient merchants to flourish as never before.

Many Mercantalists had also believed that a nation's wealth could be measured by its gold, a natural assumption on behalf of a trader. However, a landowner in a largely agricultural economy is liable to see his land as his wealth. And this was precisely the view of the Physiocrats: their third core concept. According to Quesnay, 'Agriculture is the source of all the wealth of the State and of the wealth of all the citizens.' Other forms of industry, such as those practised by merchants and traders, added nothing.

This somewhat rustic view of commerce was reflected in the

Physiocrat view of social class, which they saw as a three-tiered structure. At the top level were those who owned the land and were responsible for agricultural production. Next came the 'productive class', who actually did the work on the farm. After they had been paid enough to live on, the remaining money went to the landowners. This was then used to finance further production: next year's crop and so forth. Quesnay was against the accumulation of wealth in the form of saving. He saw that this contributed nothing. The unused gold, the very wealth of the merchants, was actually harmful to production.

Bottom of the social structure was the third class, which included merchants, traders, manufacturers, artisans and the like. These were occupations pursued in towns, rather than in the country, where natural law prevailed more ostensibly. ('Physiocrat' actually means 'rule by nature'.) According to Quesnay, such urban occupations made no contribution whatsoever to the national wealth. He even went so far as to call them the 'sterile class' (no mean insult to a Frenchman). The activities of this class were indisputably secondary. Before you could have more garment-makers, you first had to have more sheep. However, this rather jaundiced view of town-dwellers and their business left Quesnay in a quandary when it came to the vexed question of taxes. Obviously these ought to be kept to a minimum. And since, according to laisser-faire, government should be limited to maintaining law and order, and the defence of the realm, its need for revenue was suitably diminished. Even so, the landowners couldn't be expected to shoulder the entire tax bill. Yet who else was there? The 'productive class' of yokels obviously couldn't be expected to pay. They weren't paid enough in the first place. And if they *were* paid enough to be able to pay taxes, this meant the landowner would in effect be paying taxes twice over: an unthinkable state of affairs. This left the third class of urban manufacturers and their ilk. But if this despised group didn't actually produce any wealth (as had been so cogently argued), how could they be expected to pay taxes? The Physiocrats had painted themselves into a corner, and never really came up with a satisfactory answer to this question.

A close examination of Quesnay's ideas reveals a curiously personal flavour to his thinking. He came to economics in his later years, after a lifetime in medicine. His recipe for the economy bears a striking resemblance to his attitude towards health. Natural law echoes his belief in the body's ability to heal itself; laisser-faire parallels the hands-off approach. And his virulent anti-merchant views? One can't help feeling that his father's success as a small trader, and perhaps his laisser-faire attitude towards his son's education, may have had something to do with this. An ingenious French psychologist has come up with an even more perceptive suggestion. Apparently, Quesnay's condemnation of his father to the sterile class contains the implicit suggestion that the young Quesnay suspected he was illegitimate.

It was Quesnay's rational ability to perceive an overall scheme of things which was to prove most productive. The Physiocrats exhibited a curious blend of liberalism (laisser-faire) and feudalism (class system). Yet here was the first structured thinking on this subject, and some of the concepts they came up with were to remain permanent features on the economic scene. Most notable of these was of course laisser-faire, along with the first inklings of its advantages and how this worked.

Quesnay attracted several talented Physiocrat followers. The most influential of these was the politician Pierre Samuel du Pont de Nemours, who edited a collection of Quesnay's writings which he called *La Physiocratie*. This was in fact the first use of the name, and the work emphasized the central thesis that all wealth came 'from the soil'. Du Pont was to distinguish himself by writing the first history of economics, before this subject had even properly begun. The French also claim him as the first economic thinker to put his ideas into practice in government. (This national lacuna where John Law is concerned is surely understandable.) Du Pont was responsible for the free-trade treaty between Britain and

France in 1783. As a member of Louis XVI's government du Pont was a faithful royalist, but he was also strongly in favour of reform. This left him in a tricky position when the French Revolution broke out in 1789. For a few difficult years he did his best to ride the storm, offering economic advice to the revolutionaries, but he was eventually forced into hiding. In the aftermath of the Terror he was imprisoned and lucky to escape with his life. In 1799 he emigrated to America with his son Eleuthère, who had been a pupil of Lavoisier, the founder of modern chemistry. Two years later Eleuthère used his scientific expertise to set up a gunpowder factory at Brandywine Creek, near Wilmington, in Delaware. This was the beginning of du Pont, the American chemical conglomerate, which was to become one of the largest corporations in the country. Unlike similar organizations founded later in the century by the likes of Rockefeller and J. P. Morgan, du Pont would remain in the control of the family for over 150 years. The du Ponts, the Rockefellers and the Morgans would all make their fortunes in a market which practised a Physiocratic lack of restraint and regulation. Even their activities in oil, chemicals and railroads could loosely be described as making money 'from the soil'. And their contempt for the 'sterile class' (their employees) was ostensibly in accord with the notions of the Physiocrats. However, it is doubtful whether these pioneer French economic thinkers would have recognized the American robber barons as in any way the true inheritors of their system.

4

The Founding Father

Adam Smith is generally regarded as the founding father of classical economics. The title of his masterpiece *An Inquiry into the Nature and Causes of the Wealth of Nations* is in part self-explanatory. Its analysis of the economic impulses and structure of society explains how commerce works. His legacy is our understanding of the economic world we live in today.

Little is known of Adam Smith's personal life, for the simple reason that he didn't really have one. Or rather, his personal life was conducted largely within the confines of his own mind. His fellow Scot and friend, the philosopher David Hume, describes him in a letter of introduction: 'You will find him a man of true merit, though perhaps his sedentary reclusive life may have hurt his air and appearance as a man of the world.' Smith was notoriously absent-minded. (Anecdotes describe him making tea with a crust of bread, lost in hour-long reveries frowning at the sky on street corners, and so forth.) But his perception of the world he inhabited so distantly was exceptional, his knowledge of it and its ways was encyclopedic, and his acumen concerning the behaviour of both the individual and society make him one of the forerunners of modern psychology. He did not withdraw from the world, he stepped back from it in order to see it in perspective.

Adam Smith was born in 1723 at Kirkcaldy, then a small fishing village on the Firth of Forth. (Half a century previously John Law had grown up just across the water.) Smith's father had been Comptroller of Customs at Kirkcaldy and was by accounts more a 'man of parts' than this comparatively insignificant post implies.

But Adam Smith's father died several months before the birth of his famous son (an unusual circumstance which Smith shared with Newton). During his childhood Smith became inordinately attached to his mother, the daughter of a local landowner. This bond was to remain close throughout his life: he would never marry, and is not known to have formed intimate relationships with members of either sex.

At fourteen Smith went to Glasgow University, then little more than a secondary school. But the city was witnessing the beginnings of the Scottish Enlightenment, whose flowering would for a time establish Edinburgh and Glasgow as intellectual centres almost on a par with London. At Glasgow University Smith fell under the spell of Francis Hutcheson, who was not only a charismatic lecturer but also a philosopher in his own right. Hutcheson was the first professor at Glasgow to abandon lecturing in Latin. He was also a man of comparatively enlightened moral and political views. It was he who coined the famous phrase 'the greatest happiness of the greatest number', a concept which was to play a defining role in the moral philosophy of Utilitarianism in the following century. Hutcheson also believed that we are born with an inner moral sense, which is similar to our external physical ones such as sight and touch. It was only this which enabled us to restrain our more natural instincts towards self-gratification. Even in these early years Smith didn't fully agree with Hutcheson, but the perspicacity of his teacher caused him to realize that there was no easy answer to such questions. The prevailing religious and metaphysical explanations were inadequate when it came to the psychology of actual behaviour.

In 1740 Smith won a scholarship to Oxford. The seventeen-year-old student travelled 400 miles south on horseback to take up residence at Balliol College, which was then almost 500 years old and one of Oxford's most prestigious colleges. To Smith's intense disappointment, he found himself in a moribund educational backwater. Unlike his brilliant and stimulating lecturers at Glasgow, he found that in Oxford the 'professors have, for these many years, given up altogether even the pretence of teaching'. Lecturers were paid regardless of whether they lectured, and science simply wasn't taught at all. To make matters worse, there was also a prevailing 'anti-Scotch' prejudice, amongst both the students and the authorities. This presumably came to a head in 1745, during the Jacobite Rebellion. On this occasion the Scots rebels successfully invaded England, marching 200 miles south as far as Derby, before returning home because no one arrived to fight them.

At Oxford Smith made few friends and embarked upon intense bouts of self-education, interrupted by periods of illness and lassitude. In his regular letters to his mother he complained of 'an inveterate scurvy and shaking in the head'. His mother suggested drinking tar-water, which had been recommended by the Irish philosopher Berkeley as a panacea in his latest work. (Such was Berkeley's enthusiasm for this harmless tincture that philosophy had taken second place to the extolling of its miraculous virtues.) The cure proved ineffective, and Smith would suffer from 'shaking in the head' for the rest of his days. Smith's health appears never to have been of the best; but his solitude and his lack of healthy distractions ensured that he read prodigiously. The authorities considered this most unorthodox, and he was almost expelled when it was discovered that he was reading Hume's *Treatise of Human Nature*. This work was not only several centuries in advance of the philosophy then being taught at Oxford, it also contained many subversively refreshing ideas. Hume insisted that all true knowledge was based upon experience. His view that there was no room for metaphysical knowledge, in this empiricist view of the world, was to have a profound effect upon Smith. The world could only be explained by adopting a truly scientific attitude. Hume's approach was Smith's starting point.

In 1746 Smith made his way back to mother in Kirkcaldy, where he took up 'sea-bathing in the Firth of Forth summer and winter'.

During this period such behaviour was unusual, to say the least. However, his biographer Ian Simpson Ross suggests that he took up this pursuit for health reasons, rather than as an expression of eccentricity. Frigid sea water evidently reached the parts that even tar-water could not reach, for Smith continued sea-bathing whenever he stayed at Kirkcaldy. In between providing seaside entertainment for the locals, Smith embarked upon sporadic vain attempts to find employment. Eventually mother used her family connections and Smith was employed for a series of public lectures at Edinburgh. Smith's lectures would begin somewhat hesitantly, but then he would become carried away with himself and address the ceiling with intent enthusiasm for hours on end. Despite this apparent indifference to his audience, his lectures soon began attracting wide attention. The topics on which he spoke ranged through literature, jurisprudence and all branches of philosophy, with each enlightened by an exceptional erudition. Here was someone who appeared to know almost everything; but even more impressively, he had thought about what he knew. At the unusually early age of twenty-seven Smith was offered the professorship of logic at his old university in Glasgow. A year later he was promoted to professor of moral philosophy. (Despite Smith's warm recommendation, Hume was turned down for the vacant chair of logic.) In those days moral philosophy included everything from theology to ethics, from political economy to jurisprudence. Smith was in his element.

There are surprisingly few portraits of Smith: the only reliable likeness is a paste medallion which dates from later in his life. This depicts his head in profile, clad in the customary short wig of the period. He has a surprisingly hard face, with a prominent nose and protruding heavy-lidded eyes. Yet when distracted from his scowling reveries he was said to break into a smile of 'inexpressible benignity'. His shaking head, his notorious absent-mindedness and his 'vermicular' walk: such harmless eccentricities quickly endeared him to the students. Yet this apparent other-worldliness

*Adam Smith, generally regarded as the
founding father of economic theory*

contrasted sharply with the realism of Smith's social and philosophical understanding. The eighteenth century world was one of
considerable progress, but also one of brutal hardship and little
social justice. Smith regarded this world as an intellectual puzzle,
the unravelling of which was a moral duty. Humanity would only
be able to free itself from its bonds when it understood how it
was bound by them. Progress and enlightenment appeared to be
inextricably linked – but how?

Rising early, Smith would deliver his first lecture of the day at
7.30 a.m. (well before dawn in the Scottish winter). The afternoons
were spent reading and writing. Occasionally his professorial duties

93

involved him in university administration, a task for which he was ill fitted in the extreme. The man who would understand the intricacies of international commerce was all but incapable of the simplest transactions himself. Even his horse would have starved but for a friend buying its oats.

On the evenings when he wasn't reading Smith could be seen amongst Glasgow society. Here the combination of the Scottish Enlightenment and provincial smallness produced stimulating company. Literary, scientific, aristocratic and professional circles all overlapped. Since the union with England half a century previously, Scottish merchants had embarked successfully in the colonial trade, especially with the tobacco plantations of Virginia, transforming Glasgow into a thriving port. The 'tobacco lords' with their scarlet cloaks, powdered wigs and gold-headed canes mingled with the leading intellectuals of Scotland. The latter included the likes of Joseph Black, the pioneer chemist who preceded Lavoisier in the discovery of carbon dioxide; James Watt, the inventor of the steam engine; the philosopher David Hume; Dr Johnson's biographer, James Boswell; as well as the merchant Andrew Cochrane, who founded the world's first Political Economy Club. Informed conversation was liable to range from trade with the Americas to the first modern history of England, which was being written by Hume; from philosophy to the latest parliamentary issues. Hume spent most of his time in Edinburgh, just forty miles away, but he and Smith soon became close friends and correspondents. They didn't always agree. Their arguments in conversation appear to have been conducted with the vigour and wit one would expect of two such determinedly independent minds. Sadly, little of this survives in their correspondence, which consists for the most part of affectionate gossip. As Hume put it, with characteristic irony, 'I am positive you are wrong in many of your Speculations, especially when you have the Misfortune to disagree with me.' Both were writing too hard to be bothered with long-winded repetition in letters. Hume was progressing with his

History at the rate of a century a year, and Smith was composing his first major work.

The *Theory of Moral Sentiments* examines individual human nature. It is as if Smith is here preparing the way for his later work, set on the larger social stage, where this individual participates collectively in economic development. The work addresses itself to the two central questions of ethics. First, 'wherein does virtue consist?' Second, 'by what means does it come to pass, that the mind prefers one tenour of conduct to another?' To summarize Smith's long and often wayward argument: within each individual there is an 'inner man', who acts as an 'impartial spectator' to all our actions. We find his condemnation (or approval) of our actions impossible to ignore. We can defy his censure, but we cannot avoid hearing his voice. Smith's theory explains the voice of conscience, the unprompted inner sense of guilt we experience when we have done something of which we know at the same time we disapprove. It explains this contradictoriness in our nature, when there appear to be two separate elements warring within us. Smith showed how we can become overwhelmed with remorse, to the point of suicide, even when no external agency condemns us. Smith's picture of our 'inner man' is not only ingenious but also uncannily prescient. It is not difficult to see here the ghost of Freud's superego, with its socializing restraint on our instincts, which would not appear for another century and a half.

Although this was not a work on economics, it is here that Smith first introduces the idea which would guarantee him economic immortality. It is worth quoting at length:

> the proud and unfeeling landlord views his extensive fields, and without a thought for the wants of his brethren, in imagination consumes himself the whole harvest that grows upon them ... The capacity of his stomach bears no proportion to the immensity of his desires, and will receive no more than that of the meanest peasant. The rest he is obliged to distribute among

those, who prepare, in the nicest manner, that little which he himself makes use of . . . The rich only select from the heap what is most precious and agreeable. They consume little more than the poor, and in spite of their natural selfishness and rapacity, though they mean only their own conveniency, though the sole end they propose from the labours of all the thousands whom they employ, be the gratification of their own vain and insatiable desires, they divide with the poor the produce of all their improvements. They are led by an invisible hand to make nearly the same distribution of the necessaries of life, which would have been made, had the earth been divided into equal portions among all its inhabitants, and thus without intending it, without knowing it, advance the interest of the society.

Economics had found its voice. And also, in the 'invisible hand', its guiding principle. But this can only be seen with hindsight. In *The Theory of Moral Sentiments* Smith was still concentrating upon how socializing forces play upon the individual, rather than the larger scheme of things.

Adam Smith's first major work was published early in 1759. Hume, whose great *Treatise of Human Nature* had been published twenty years earlier to almost complete public indifference, had warned his friend to accept such a fate philosophically: a true philosopher will always have his hopes dashed. Within the month Hume was writing wryly to console his friend from London:

Supposing . . . that you have duly prepared yourself for the worst . . . I proceed to tell you the melancholy News that your Book has been very unfortunate: For the Public seem disposed to applaud it extremely. It was look'd for by the foolish People with some Impatience; and the Mob of Literati are beginning already to be very loud in its Praises.

Adam Smith was famous, his book an overnight sensation.

*

One of the many who read Smith's book was the statesman Charles Townshend, a man whose exceptional financial acumen was counterbalanced by his political judgement. During his period as chancellor of the exchequer he would almost single-handedly succeed in sparking the American Revolution (it was his taxation policy which resulted in the Boston Tea Party). But Townshend's financial ability meant that his judgement was somewhat more assured when it came to his personal life. He had recently married the widow of the Duke of Buccleuth, thus allying himself to one of Scotland's richest and most influential families. Townshend was so impressed by *The Theory of Moral Sentiments* that he wrote inviting its author to become personal tutor to his teenage stepson, the new Duke of Buccleuth. His terms were exceedingly generous: £300 a year during his period as tutor, followed by a pension of £300 a year for life. (As a professor at Glasgow, Smith was earning just £200 a year.) At the age of forty Smith resigned his post and set off with his young charge on the customary educational grand tour of Europe. However, for over a year Smith found himself stuck in Toulouse. His French was atrocious and he found the local provincial society incurably boring. During the sweltering July of 1764, Smith wrote to Hume, 'I have begun to write a book in order to pass away the time.' This was the beginning of Smith's greatest work, which would take him another twelve years to complete. Its international wealth of reference would owe not a little to his first-hand experiences on his tour of Europe.

After Toulouse, Smith and his aristocratic charge (now joined by the duke's younger brother) travelled to Geneva. Here, Smith was able to meet Voltaire, at the time the most famous writer in Europe. Voltaire was by now 'cultivating his garden' at Ferney, his country estate above Lake Geneva. His enlightened wit and liberal ideas had brought him into constant conflict with the despotic French *ancien régime*, and also into occasional ructions with the Swiss authorities. In order to overcome these difficulties he lived in an estate which stretched either side of the Swiss–French border, thus ensuring that he could swiftly retire from the offended jurisdiction. Smith recognized at once that Voltaire was more than just the provocative, clever author he appeared to many. They met 'five or six times', conversing at length, and quickly became close friends. Though, as with Hume, the actual cut and thrust of these conversations which so endeared them to one another remains unknown. All we know for certain is that, as usual, Voltaire ventured his opinion that English cuisine possessed only one sauce (melted butter). In later life, Smith would say of Voltaire, 'Reason owes him incalculable obligations.' Politics, philosophy, science: all were now making giant strides with the aid of reason. The time was ripe for the embryo economics to take a similar step.

In 1765 Smith and his two charges arrived in Paris. Here, Hume was now secretary to the British Embassy, and ensured that his friend was introduced to the leading salons of the French Enlightenment. Smith became particularly friendly with the mathematician and philosopher d'Alembert, who was devoted to the rationalist aim of uniting all branches of knowledge. In characteristic French fashion d'Alembert even established 'reasoned principles' by which the various sciences were linked. D'Alembert, along with the philosopher Diderot, was also the moving spirit behind the great French *Encyclopédie*, whose aim was the propagation of the new sciences, liberal ideas and secular thinking. D'Alembert's belief in the interconnection between forms of knowledge would play a leading role in Smith's thinking. In his new book his developing economic ideas would provide an intimate link between philosophy, psychology, history and science, including politics, commerce, sociology and even mathematics.

Such would be the original branch of knowledge which Smith's coming work would establish. But his central ideas would not be entirely original. Adam Smith's notion of economics would come to rely heavily upon the ideas of the Physiocrats, whose leading members he now encountered in the salons of Paris. Quesnay was at the time Louis XV's physician, and informal discussions amongst the Physiocrats were regularly held in Quesnay's apartments at the royal residence. Smith was often invited to join these, though according to du Pont the contributions made by the Scots visitor were not highly regarded. How much of this was due to Smith's murderous treatment of the French language is unclear. Also, Smith's ideas were still in the embryo stage, while the Physiocrats had a highly inflated notion of their own importance in the history of intellectual progress. In 1758 Quesnay had published a *Tableau économique* (economic table) which showed in diagramatic form how money flowed between the different classes (it was here that the *classe sterile* made its first notorious appearance). Quesnay's *Tableau économique* was soon being hailed as a discovery on a par with the invention of printing or the first use of money. Despite such hyperbole, it is not difficult to see how Smith's 'invisible hand' in *The Theory of Moral Sentiments* now developed into something much closer to the Physiocrats' all-pervasive *droit naturel* (natural law). Smith certainly recognized this debt, and it was his intention to dedicate the new book he was writing to Quesnay, but the Frenchman would die two years before its eventual publication.

Smith's stay in Paris was dramatically cut short in October 1766. According to one source, the Duke of Buccleuth's young brother, the Honourable Hew Scott, 'was assassinated in the streets of Paris'. However, from Smith's frantic letters to Lady Frances Scott, the sister, it appears that the nineteen-year-old was in fact struck down by a sudden and violent fever. Smith summoned 'my particular and intimate friend Quênai [sic] . . . one of the best Physicians that is to be met with in any country'. (He succeeds in spelling Quesnay's name three different ways in the same letter, each of them more ingeniously wrong.) After the manner of the period, Quesnay prescribed 'opiat' and 'fifteen grams of Rhubarb', but to little avail. The patient was soon raving, and within a few days he died.

Smith returned to London with the distraught young Duke of

Buccleuth. Townshend was remarkably understanding about the whole affair, and no blame was cast upon the person who had been hired to look after his stepsons' welfare. Smith remained in London for a while, during which he was elected a Fellow of the Royal Society. Then, at the age of forty-four, he returned north of the border to stay with his mother at Kirkcaldy. Here, he lived off his generous pension from Townshend, and devoted himself to completing the book he had started in France. This proved an exactingly complex task, which became a great drain on Smith's mental and physical resources. After two years he explained in a letter, 'my own schemes of study leave me very little leisure . . . I scarce see any probability of their ending'. Two years later he was writing, 'I shall not leave my retreat for above a day these six months to come.' Smith found writing difficult; but he had also set himself a mammoth task. This project was nothing less than laying out the scientific foundations of an entire new field of knowledge. Hume missed him and wrote from Edinburgh, just twelve miles across the water with the odd islet in between: 'There is no Habitation on the island of Inch-keith otherwise I shoud challenge you to meet me on that Spot, and neither us ever to leave the Place, till we have fully agreed on all points of Controversy.'

Smith was to remain at Kirkcaldy for six years, during which time his eccentricity appears to have come into its own. From this period dates the celebrated anecdote concerning one Sunday morning in Dunfermline, when the bells were ringing and the locals dressed in their Sunday best were making their way to church. Smith suddenly appeared in the street clad only in his dressing gown. His loyal biographer Ross ingeniously tries to explain this away: 'He may have walked the fifteen miles from Kirkcaldy in a fit of abstraction, perhaps after taking a wrong turning after a douse in the North Sea.' Admittedly, this would seem to be as likely an explanation as any.

After his six years at Kirkcaldy Smith moved to London, but

still he wasn't finished with his book. It would be another three years before he finally published *An Inquiry into the Nature and Causes of the Wealth of Nations* in 1776, less than six months before the American Declaration of Independence. (It is easy to see the shared birthdate of America and economics as more than just historical coincidence.) Adam Smith's masterpiece runs to over 900 pages. It was justly described by the American sociologist Max Lerner as 'the outpouring not only of a great mind, but of a whole epoch'. Its breadth of reference extends from the salt money of Abyssinia to the smuggling of French wine into Britain from Dutch Zealand. It is filled with perceptive observations: 'Science is the great antidote to the poison of enthusiasm and superstition,' and 'with the greater part of rich people, the chief enjoyment of riches consists in the parade of riches'. Smith was particularly interested in riches and money. He even went to great lengths to calculate that for each course of his lectures the ancient Greek orator Isocrates made £3,333.6s.8d (probably the equivalent nowadays of a cabinet minister's salary).

It would be over a century before this cornucopia of multifarious knowledge received the index it deserved, with Edwin Cannan's edition in 1904. This index, a classic in its own field, extends over sixty tightly printed pages, and contains such idiosyncratic antiquities as: 'Clergy, must be managed without violence'; 'Idleness, unfashionable in Holland'; 'Ireland, supplies strong porters and beautiful prostitutes, fed on potatoes, to London'; 'Siberia, barbarous because inland' and 'Venison, the price of, in Britain, does not compensate the expence of a deer park'.

Smith's interest in the expense of maintaining protected estates populated by enthusiastic hunters also extended to larger enterprises than deer parks. He noted that the cost of the civil government of the Massachusetts Bay colony was about £18,000 pa, whereas that of Connecticut was a mere £4,000. Smith specifically mentions these figures as being 'before the commencement of the present disturbances', though a few pages on he calls these 'the late disturbances'. He evidently surmised that the American Revolution would have come to an end before his book was finally published.

Yet this gaffe is a rarity. Smith was alone in the eighteenth century in sensing the future greatness of America:

> ... in the course of little more than a century, perhaps, the produce of America might exceed that of British taxation. The seat of the empire would then naturally remove itself to that part of the empire which contributed most to the general defence and support of the whole.

No one else even dreamt that such a shift of power might take place. The French thought their American colony of so little importance that in 1803 they sold the Louisiana Territory to the United States for a mere 3 cents an acre. The United States doubled its territory at a stroke, in the greatest real estate bargain of all time. (The Physiocrat du Pont was one of the leading negotiators of this deal, acting on behalf of the French – thus inadvertently contributing far more to America than the mere industrial giant established by his son.) Ironically, amongst all the great nineteenth-century thinkers it would be a Frenchman who understood the truth of what Adam Smith had foreseen in the previous century. De Tocqueville alone predicted America's greatness: a detail which was overlooked by superior thinkers who had so much else to say about the twentieth century – such as Hegel, Marx, Nietzsche, et al.

All this richness of detail and digression means that *The Wealth of Nations* is open to widespread interpretation. Despite this, Smith's central picture remains clear enough. His second major work extends the philosophy first introduced in *The Theory of Moral Sentiments*. The psychological drama involving the passions and the 'impartial spectator' is now extended to the wider stage of society, where it becomes a social struggle. Smith realizes that in order to understand this social struggle properly it is first necessary to trace its evolution through history (though curiously, he leaves his detailed examination of this to the end of the book, and various lectures). Unless this evolution is hampered by bad government, war or lack of resources, society will progress through four distinct historical stages, each of which will generate its own appropriate mores and institutions.

The first stage is the 'lowest and rudest, such as we find it among the native tribes of North America' and in England before the invasion of Julius Caesar. Society consists of small tribal groups, where 'every man is a warrior as well as a hunter'. 'There is scarce any property,' which means disputes are rare. 'So there is seldom any established magistrate or regular administration of justice.' The absence of property, lack of authority and laws, together with characteristic activities such as hunting and inter-tribal skirmishes, all encourage self-reliance and allow a greater level of personal liberty.

The second stage is 'nations of shepherds . . . such as we find it among the Tartars and Arabs'. This kind of nation is nomadic. 'Whether it marches as an army, or moves about as a company of herdsmen, the way of life is nearly the same.' The common pastimes of these people are 'running, wrestling, cudgel-playing, throwing the javelin, drawing the bow . . . all of them the images of war'. As a result, when they are united under a leader like Ghengis Khan or Mohammed, the effect can be devastation on an inter-continental scale. These people fight for no recompense, seeking only to gain plunder. 'Nothing can be more dreadful than a Tartar invasion.' Smith warns, 'If the hunting nations of America should ever become shepherds, their neighbourhood would become much more dangerous to the European colonies than it is at present.'

Yet keeping flocks required a more advanced social organization than amongst the tribal Native Americans. The nomadic shepherd societies had private property, which had to be protected by law. This central institution introduced a formalized state of law and order. Here lay the essence of society as we know it. But Smith discerned that this also introduced a basic inequality. 'Civil government, so far as it is instituted for the security of property, is in reality instituted for the defence of the rich against the poor, or of those who have some property against those who have none at all.'

Smith then demonstrates how society evolved through the stage leading from ancient Greece and the Roman Empire, through to feudalism. Such agriculture-based societies required armies that had to be supported by the state. 'The number of those who can go to war, in proportion to the whole number of the people, is necessarily much smaller, in a civilized, than in a rude state of society.' This partly explained why the primitive Visigoths and the Tartars managed to overcome more advanced societies in Europe and Asia. As society progressed, it became less warlike. Shepherds have a great deal of leisure, farmers less so, and workers practically none at all. 'The first may, without any loss, employ a great deal of his time in martial exercises,' whereas 'the last cannot employ a single hour in them without some loss'. Wages, production, inequality, ownership: all these became more accentuated – to the point where some members of society actually owned others, in the form of slaves, vassals or serfs. As civilization advanced, so its materials and means of production developed, bringing about an evolution in the hierarchy, institutional development and new laws. In the following century, Marx would develop such thinking into an overall theory of history. Yet the difference between him and Smith remains fundamental. For Smith the driving force behind history was 'human nature', which was always motivated by self-interest. For Marx, it was the conflict between classes. One society consisted of individuals, the other of groups: a telling distinction.

In the medieval world, wages and production were controlled by the merchant guilds, the market controlled by the government to protect its merchants. But now, in the fourth and final stage of society, the market itself is the controlling force. When this is unrestrained, it is every man for himself. 'It is not from the benevolence of the butcher, the brewer, or the baker, that we expect our dinner, but from their regard to their own self-interest. We address ourselves, not to their humanity but to their self-love.' From Smith's hearty dinner of chops, ale and bread we discover the entire mechanism of society.

> As every individual, therefore, endeavours as much as he can both to employ his capital in the support of domestick industry, and so to direct that industry . . . in such a manner as its produce may be of the greatest value, he intends only his own gain, and he is in this, as in many other cases, led by an invisible hand to promote an end which was no part of his intention.

Here again, we come to the heart of the matter. Just as it did for the avaricious landlord in *The Theory of Moral Sentiments*, this same invisible hand guides the self-interested manufacturer to promote the public good. Such a pessimistic analysis of human behaviour has recognizable echoes of Mandeville's 'private vices, public virtues'. But they are not the same. For Mandeville this was an insight; for Smith, self-interest becomes the driving force behind all social development, and every individual involved in it.

The invisible hand is Smith's most memorable and evocative image, though surprisingly this central image in fact occurs just once in each of his great works. Yet, contrary to misguided belief, it has no theological overtones. As J. K. Galbraith pointed out, 'A man of the Enlightenment, Smith did not resort to supernatural support for his argument.' The invisible hand was simply his understanding of the way the world worked. His intention was strictly scientific: this was to be the Newtonian gravity of the new science of economics.

But how precisely did the invisible hand work? In *The Theory of Moral Sentiments* the inner 'impartial spectator' restrained the natural passions. Moving out on to the larger stage of society, we find individual self-interest restrained by competition. It is this which regulates the selfish desire to better ourselves, 'a desire that comes with us from the womb, and never leaves us until we go into the grave'. It was competition between the Glasgow butchers, competition between the brewers, and competition between the bakers which dictated the price of the lamb chops, brown ale and soda bread on which Smith dined. This competition kept the prices of all commodities down to 'natural' levels. Here again, we see an echo of Physiocrat ideas with their metaphysical overtones. But the invisible hand is much more (and much less) than the 'natural law' (*droit naturel*). And the 'natural' level of prices is a similar scientifically conceived concept, stripped of metaphysics.

Smith's idea of science was based on Newtonian principles. The natural world consisted of nothing more than the mechanical motion of matter. Motion was passed on, and in this way preserved. Thus nature's objective was achieved. This applied both to objects and human beings. But unlike objects, human beings were preserved by passing on life. Hence the most natural way for the economic world to operate was a mechanical system which organized things in the best way to sustain life. This was achieved by the system which we now know as the law of supply and demand. Basically, prices increase as demand increases; they decrease as supply increases. This was the 'natural' law which determined the level of prices, and thus the level of wages and profits.

There are many huge jumps in this argument which are not supported by logic, or even by evidence – despite Smith's abundance of economic lore. But its force is undeniable. What it discerns is a fundamental principle. This is what happens in an unrestrained market. Objections that Smith's argument is not logically deduced, or scientifically supported, are beside the point. Here was a concept which worked: it could be used. Despite protests to the contrary, even scientists of the first water proceed in just this fashion: the atom proved a highly fruitful idea for over two centuries before even a glimmer of scientific evidence was found for its existence.

However, this is only true for the outline of the mechanism which Smith discerned at work. When it came to details, there were not so much gaps, but holes in the argument. Again, this is understandable: this was, after all, just the beginning of a completely new branch of knowledge – or science, as many would have it. These 'holes' in Smith's argument were topics which were to bedevil economics for over two centuries to come. For instance, the 'natural' price determined wages, rent and profits. But how was this naturally determined sum to be divided between these three? Obviously, the less a manufacturer paid in wages, the more he gained in profit. Once again, the law of supply and demand operated. With workers available in abundance, it was possible to pay lower wages. But there was a lower limit. No worker would, or could, work for wages below the level of subsistence. In this brutal new economic world wages seemed 'naturally' to gravitate towards subsistence level.

Smith was a humane man, and was appalled at what he appeared to have 'discovered'. Fortunately, he was quick to see the fallacy in what came to be known as the 'subsistence theory of wages'. In order to maintain a ready supply of workers, wages had to be high enough for a worker to marry and maintain a healthy family. If a nation was to preserve its wealth, the future had to be taken into account. Subsistence wages were not enough. The elevation of this subsistence fallacy to a 'theory' would remain a moral bugbear. As workers became more mobile, they became more expendable – a lesson all too clear to the dispossessed rural poor of Smith's day, as it is to the migrant Third World worker of today.

Smith was no sentimentalist: he sought to discover his new science in the world as it is, not as it ought to be. Yet he seldom lost sight of the fact that this was a science of human lives. Human nature may be driven by naked self-interest, but this could be modified by the 'impartial spectator' within us. Yet what motivated this neutral onlooker? In Smith's view, reason and sympathy were his guiding principles. This could perhaps encourage some employers to take less profit than was available under the 'natural law'.

But back to the real world. Smith observed that there were two separate forms of value. A product had a 'value in use' and a 'value in exchange'. But this gave rise to a puzzling anomaly. 'Things which have the greatest value in use have frequently little or no value in exchange.' And vice versa. For instance, 'Nothing is more useful than water: but . . . scarce any thing can be had in exchange for it.' By contrast diamonds, 'the greatest of superfluities', have 'scarce any value in use', but can be exchanged for huge amounts. Smith never came up with a completely satisfactory answer to this problem. Or rather, he came up with a number of vague suggestions which subsequent economists would interpret in their own fashion. Indeed, their answers to this problem would frequently provide the moral base of their system. Essentially, Smith decided to ignore a product's value in use. What mattered was its value in exchange. But what precisely did this mean? What, ultimately was value in exchange? 'The value of any commodity . . . to the person who possesses it . . . is equal to the quantity of labour which it enables him to purchase or command. Labour, therefore, is the real measure of the exchangeable value of all commodities.' This would become known as the Labour Theory of Value.

As Smith emphasizes from the start, wealth is increased by 'the skill, dexterity, and judgment with which labour is applied in any nation', and also by 'the proportion of those who are employed in useful labour, and that of those who are not so employed'. In other words, economic growth is promoted by increasing productivity. This is best achieved by the division of labour. Smith takes over Petty's phrase; but, as with his other borrowings, Smith's analysis goes far beyond the original. His celebrated example is a pin factory. This employs ten workers, each specializing in a separate task.

> One man draws out the wire, another straights it, a third cuts it, a fourth points it, a fifth grinds it at the top for receiving a head;

to make the head requires two or three distinct operations; to put it on, is a peculiar business, to whiten the pins is another; it is even a trade by itself to put them into the paper . . .

In this way ten persons could make over 48,000 pins in a day. But if they had all worked separately and independently, each one would have been able to produce less than twenty a day, if that.

This homely example is characteristic of the man and his era. Smith's emphasis is on skills, rather than machinery. As Galbraith perceptively remarks, 'Adam Smith gave economics its modern structure. But that structure was given him, in turn, by the earliest stages of the Industrial Revolution.'

Smith's description of the pin factory shows all the hallmarks of active curiosity (though treacherous scholarship has revealed that this description was in fact 'gleaned' from a work by the French philosopher Diderot). Yet Smith's wealth of observation is more than just facts taken from books. Many of his details indicate precise observation and widespread experience. The strength of Irish porters; the beauty of potato-fed prostitutes on the streets of London; how in France 'the labouring poor seldom eat butcher's meat'; Voltaire's opinion that 'Porée, a jesuit of no great eminence in the republic of letters, was the only professor they had ever had in France whose works were worth reading' – these were things Smith had seen and heard. He may not himself have observed in minute detail the dozen or so separate operations required to make a single pin, but he had observed the effect these minute, repetitive operations had on the people who undertook them. Dividing labour wreaked its effect on the labourer. He had 'no occasion to exert his understanding', and thus lost this habit, becoming 'as stupid and ignorant as it is possible for a human creature to become'. All Smith can do is accept the iron laws of this new science: 'in every improved and civilized society this is the state into which the labouring poor, that is, the great body of the people, must necessarily fall, unless government takes some pains to prevent it'. Smith makes no suggestions at this point: he sees himself as a scientist, not a politician.

Yet the persistent force of his description and analysis make it plain that he did not like what he saw. Even the poorest inhabitant of a village felt obliged to behave himself, because he was constantly observed by his neighbours. But 'as soon as he comes into a great city, he is sunk into obscurity and darkness. His conduct is observed and attended by nobody and he is therefore very likely to neglect himself, and to abandon himself to every sort of low profligacy and vice.'

Smith deals in some detail with drunkenness, a perennial British topic. However, his analysis and prescription may perhaps be a little too Scottish for some tastes. 'The inhabitants of the wine countries are in general the soberest people in Europe.' Drunkenness was only prevalent in nations where there were no grapes. That is, 'among the northern nations, and all those who live between the tropics, the negroes, for example, on the coast of Guinea'. On the other hand, 'It is not the multitude of ale-houses ... that occasions a general disposition to drunkenness among the common people.' The cause lay in the excessive cost of alcoholic drinks of all kinds.

> Were the duties upon foreign wines, and the excises upon malt [whisky], beer, and ale, to be taken away all at once, it might ... occasion in Great Britain a pretty general and temporary drunkenness among the middling and inferior ranks of people, which would probably be soon followed by a permanent and almost universal sobriety.

A likely story. Admittedly, such rash optimism is rare in Smith – which perhaps accounts for the heady effect on the few occasions when he does resort to this freely available intoxicant. Or perhaps we should put this howler down to a Scottish unwillingness to part with his cash. Either way, Smith should have known better. It was only fifty years since the almost unlimited availability of cheaply distilled spirits had devastated the poorer districts of London far, far worse than any modern crack epidemic. That had been the period of 'drunk for a penny, dead drunk for tuppence' which had inspired Hogarth's bitter *Gin Lane*. It would have been a vivid memory for anyone of the preceeding generation – but Smith drew no lesson from this, it would seem.

Smith's further analysis returns to his customary incisive level. As ever, he is perceptive of social trends. He notes that not all the dispossessed rural poor, forced to labour in the dark, satanic mills of the early Industrial Revolution, were driven to drink. Some found solace in religion, especially in the small Nonconformist Christian sects which began springing up in the poorer areas of the new cities. Here in the chapels, the dispossessed working man, cut off from his roots in a neighbourly village, would find 'a degree of consideration which he had never had before'. Such sects formed close, self-supportive communities, though their morals were too frequently 'rigorous and unsocial'. Smith's remedy for such small-minded bigotry, as for the curse of drunkenness, was education. Likewise, the sectarian bitterness engendered by such groups could be ameliorated by instilling an interest in the arts. Painting, music, dancing and theatre (especially satire) all made for a more tolerant society.

Smith's call for the education and civilization of the working classes is all very laudable. But it is in fact self-contradictory. He accepts the division of labour, which makes workers 'as stupid and ignorant as it is possible for a human being to become', as the regrettable but inevitable price of progress. The introduction of education and culture would only render workers unwilling to submit to such mind-numbing tasks. You can't have it both ways.

Yet on the whole Smith is excellent at revealing the anomalies and hypocrisies which make up society. He discerned that in every civilized society, with its class distinctions, 'there have always been two different . . . systems of morality current'. One is strict, the other liberal. 'The former is generally advanced and revered by the common people: the latter is commonly more esteemed or adopted by what are called people of fashion.' In the liberal system luxury, 'wanton and even disorderly mirth', pleasure, intemperance, 'the breach of chastity . . . are easily either excused or pardoned altogether'. For the strict system such excesses are viewed with 'abhorrence and detestation . . . The vices of levity are always ruinous to the common people, and a single week's thoughtlessness and dissipation is often sufficient to undo a poor workman for ever.' Modern social conditions are less harsh on 'levity', and the line between liberal and strict is now more blurred. Yet the ghost of this hypocritical divide lives on, its spectral nuances still discernible in the pronouncements of politicians and tabloid headlines.

All this may sound more sociology than economics to the contemporary ear. Yet the force of Smith's economic arguments gain their strength from the reality of his social observation. He spoke of actual people in an actual, historically recognizable country. These were not the mythological or mathematical races which would appear in the fantasies of future tyrannical systems, either social or economic (the two are not always the same). As society underwent the upheaval of the early Industrial Revolution, observers such as Smith began to see with new eyes the world they were living in, and how it worked. Previously, the intellect had tended to regard the human condition as a philosophical or theological matter: an affair of the soul. The material background to this condition was described by history and science. The spiritual and material worlds, mind and matter, were two separate spheres. Human life in this world took place at the tangential point where these two spheres made contact. Now it was becoming clear that there was more than just contact at this point. The human condition was also social, economic, historical. Over the ensuing centuries the two spheres – the spirit and its worldly setting – would be seen increasingly to overlap.

The invisible hand, division of labour, economic growth – our modern understanding of these terms is based on Smith, even when the idea was not originally his. The summation of Smith's economic world comes in a term that was originally his, in the sense we now use it: namely, free trade. When trade was unrestrained by tax, excise or other control, the invisible hand of the market could proceed to do its work. As we have seen, this argument ultimately derives from Mandeville's 'private vices, public virtues' (though Smith goes to some lengths to deny this). Yet surprisingly, Smith backs up his free trade/invisible hand argument with an example that flies in the face of Mandeville's prescription. According to Smith, 'It is the maxim of every prudent master of a family, never to attempt to make at home what it will cost him more to make than to buy.' He goes to the market.

> What is prudence in the conduct of every family, can scarce be folly in that of a great kingdom. If a foreign country can supply us with a commodity cheaper than we ourselves can make it, better buy it of them with some part of the produce of our own industry, employed in a way in which we have some advantage.

Here the private virtue of thrift becomes a public one. Mandeville's recipe is thus confirmed as a revealing insight, rather than a universal prescription.

The larger the free-trading area, the more room there is for every country within it to specialize in what it does best. Here is division of labour on an international scale, though a few pages later Smith admits that absolute free trade is a utopia which can never be achieved. No state can function without taxes, which are by their very nature restrictive of some element of trade. Taxes must always be raised for defence, justice and public works. Even above this there will always be protectionism of some sort, owing to 'not only the prejudices of the public, but what is much more unconquerable, the private interests of many individuals'.

Smith is explicitly against Mercantilist protectionism. But he also castigates the merchants who profit from free trade. Their acquisitive behaviour may have benefited society, but he still couldn't hide his distaste for their motives. Such greed and scheming mentality were utterly alien to his academic nature. But what did he expect? Private vices might become public virtues, but they couldn't become private virtues as well. Smith's social distaste also reveals a political fear, when he speaks of 'the mean rapacity, the monopolizing spirit of the merchants and manufacturers, who neither are, nor ought to be, the rulers of mankind'. Here we have another contradiction. Under Smith's system, such people would seem bound to become the 'rulers of mankind'. (Indicatively, this plutocratic class has now expanded beyond more politically incorrect mankind. Its exultant members today see themselves as 'masters of the universe', or on occasion 'the biggest swinging dick on Wall Street' – thus tending to both ends of the evolutionary scale.)

Smith is also against the Mercantilist notion of wealth being seen in terms of gold. 'Gold and silver are merely tools, no different from kitchen utensils ... their import increases the wealth of a country just as little as the multiplication of kitchen utensils provides more food.' The wealth of a nation grows as it learns how to compete on the free market. Despite the impossibility of absolutely free trade, Smith's ideas on the market remain a shibboleth of our modern 'free world'. Here the notion of 'free' is intentionally blurred: referring to both trade and civil liberties. We may have misgivings about the actual extent of both in this so-called free world. Indeed, in parts these are undeniably under severe restriction – by multinationals as well as governments. Yet the link between liberal trade and liberal government appears to be intimate. Western liberal capitalism, and the free democratic institutions with which it is associated, would seem to go hand in hand. They work in a fundamentally similar way: both are dependent upon freedom of choice. But like so much social binding the link between capitalism and democracy is in fact an illusion: a collective myth to which we willingly subscribe. There is no logical necessity about their concurrence. Socialist communities have often achieved great individual freedom (as did primitive

Native American societies). And democratic capitalism has on occasion failed so utterly to protect the freedom of its citizens that they have willingly voted in a dictatorship in its stead. (Hitler, along with many Third World 'life presidents', initially took office through a democratic constitutional process.) Protest that such examples involve more complex factors only confirms the point. The link between Adam Smith's free market and free democratic government is not inevitable – it is simply the fortunate belief of our age. Ominously, it would seem that such a belief is unlikely to survive into an age when global resources become increasingly limited. Here, conversely, there would seem to be an undeniable link: as the market becomes restricted, so will its associated freedoms.

Modern life is assumed to be a reflection of modern thought. As one changes, so will the other. We see 'freedom' as central to our present life, and believe this is reflected in our thought. Yet in fact we spend far more time thinking about money (many of us even equate this chameleon entity with freedom). As our life revolves more and more around money and commercial worth, so commercial value assumes an ever more central role in our thinking. Caught in this vicious circularity, we can only ask ourselves, 'Is it worth it?'

Smith had spent over a decade writing *The Wealth of Nations*. In the words of his great friend Hume, 'It was a Work of so much Expectation, by yourself, by your Friends, and by the Public, that I trembled for its Appearance.' By the time Hume received his copy, he was already on his deathbed, but he managed to write back, 'I am so much pleas'd with your Performance, and the Perusal of it has taken me from a State of great Anxiety . . . it has Depth and Solidity and Acuteness.' Though he couldn't refrain from adding, 'If you were at my Fireside, I should dispute some of your Principles.' Alas, this was not to be. Six months later Hume was dead.

Word soon spread throughout Europe of Smith's new work. It was translated into half a dozen languages; but true to form it was dismissed in Oxford as beneath academic attention. *The Wealth of Nations* was not to achieve the same popular success in Britain as his *Theory of Moral Sentiments*, yet it was read by those who mattered. The movers and shakers of eighteenth-century Britain, even future prime ministers, all took note of Smith's ideas – which were soon being put into practice.

A year after the publication of *The Wealth of Nations*, Smith was offered the post of Commissioner of Customs and Salt Duties in Scotland, in recognition of his services to the nation. Despite this being in complete contradiction to his free-trade ideas, Smith surprisingly accepted the post, which was a virtual sinecure. However, when he read the list of goods prohibited by the Customs, he discovered 'to my great astonishment, that I had scarce a stock, a cravat, a pair of ruffles, or a pocket handkerchief which was not prohibited to be worn or used in Great Britain. I wished to set an example and burnt them all.' He strongly advised his friends not to examine their own household furniture and possessions 'lest you be brought into a scrape of the same kind'.

Yet despite being an instrument of the nation's protectionism, Smith continued to advocate his free-trade views. As a theoretical expert in the hypocrisy of social behaviour, he presumably felt the need for some practice in this field.

At the age of fifty-four, together with his aged mother, Smith moved to a house on Canongate in Edinburgh. They were tended by his cousin, Miss Jane Douglas, who looked after him 'in a sisterly fashion'. Smith continued to spend most of his time writing and reading. Despite their separate subject matter and achievements, *The Theory of Moral Sentiments* and *The Wealth of Nations* were in fact intended as part of a greater overall scheme which Smith had long had in mind. As he wrote in a letter, 'I have likewise two other great works upon the anvil; the one is a sort of Philosophical History of all the different branches of Literature, of Philosophy, Poetry and Eloquence; the other is a sort of theory and History of Law and Government.' But as we have seen, Smith was a painstakingly slow writer, and consequently his ambitious project was never to be fulfilled. In 1790, at the age of sixty-seven, Smith died. In his will, he expressly ordered that all his papers be burnt.

The inner workings of the mind were echoed in the public workings of commerce. The 'impartial spectator' restrained our passions, just as the 'invisible hand' manipulated our self-interest. How might these have been translated into a philosophy of culture and a theory of law? What we know of Adam Smith is but part of the whole, but this was enough to open up an entirely new branch of human knowledge: economics. Just two years after his death, the future prime minister of Britain, William Pitt the Younger, addressed parliament with his budget speech – in which he assured the nation that Adam Smith could 'furnish the best solution of every question connected with the history of commerce and with the system of political economy'. Adam Smith had arrived centre-stage on the political-economic scene, a position he continues to occupy to this day.

5

French Optimists and British Pessimists

French economic and social thinking, upon which Adam Smith had based so many of his ideas, continued to develop independently during the latter half of the eighteenth century. One man was to have widespread social influence during this period, though his deeper intellectual significance was not to become apparent for almost a century and a half. This man was Condorcet – or Marie-Jean-Antoine-Nicholas de Caritat, Marquis de Condorcet, to give him his full title. Condorcet was born of minor aristocracy during 1743 in Ribemont, a small town eighty miles northeast of Paris. As with a surprising number of geniuses in all fields, Condorcet was brought up by a single parent. His father, a cavalry officer, was killed in action just a few days after his birth.

Condorcet quickly showed exceptional promise as a mathematician, and after rejecting his family's plans for a military career, he settled in Paris. Here he presented a paper on differential calculus to the Académie des Sciences. This was read by the great d'Alembert, who recognized its brilliance; as a result, Condorcet was elected to the Académie at the exceptional age of twenty-six. D'Alembert virtually adopted Condorcet as his protégé, and certain similarities in their upbringings appear to have inspired a deep mutual affection. St Jean le Rond d'Alembert was also the son of a minor aristocrat, and did not grow up in a conventional family. But in d'Alembert's case he had been abandoned as an infant bastard on the church steps of St Jean-le-Rond – hence his unusual first names (meaning 'St John the Round'). A leading figure of the

*The Marquis de Condorcet, a 'volcano covered
with snow'*

Enlightenment, d'Alembert soon introduced Condorcet to the intellectual circle associated with the *Encyclopédie*, whose publication was intended to propagate the new rational, scientific knowledge. Condorcet himself now began making contributions to the *Encyclopédie*. Condorcet possessed a great enthusiasm for the latest ideas, but remained nonetheless somewhat reserved by nature. D'Alembert described him as a 'volcano covered with snow'.

This was precisely the time when Adam Smith had befriended d'Alembert in Paris, and Condorcet certainly met Smith in the salons. They are said to have discussed Physiocrat ideas, and even agreed upon certain flaws in this approach, such as the emphasis upon agriculture as the source of all wealth. Despite this,

Condorcet's attitude towards economic theory was to be heavily influenced by the Physiocrats. He became a firm advocate of free trade and laisser-faire.

Mathematics, the Enlightenment, and now Physiocrat ideas on commerce – Condorcet was assembling a considerable intellectual arsenal. Yet this remained largely theoretical. Then in 1770 Condorcet was introduced to Voltaire at Fernay. Inspired by their conversations, Condorcet understood how he could develop his ideas in practical fashion. France was in need of more than just Enlightenment. The moribund structure of the *ancien régime* needed thoroughgoing reform in the social, economic and political spheres. Condorcet conceived the idea that such reform was possible to achieve in a scientific fashion.

This concern led directly to Condorcet's first major work, which was in fact a mathematical paper, whose application to economics would not be fully realized until many years after his death. In 1785 Condorcet published his *Essay on the Application of Analysis to the Probability of Majority Decisions*. This was the first attempt to apply probability theory to collective decision-making – as, for instance, in elections. Science could be applied to nature, so why not to human nature? This led to two distinct types of probability. First there was 'abstract' probability, such as involved in throwing dice: this was the absolute probability of nature. Yet where human nature was concerned there was subjective probability, which gave 'grounds for belief' that something would occur. Condorcet maintained that these were both part and parcel of the same probability process. 'A very great abstract probability gives grounds for belief that are close to certainty' (for instance, the belief that if you throw a dice 100 times, it will at least once come up 6). The less abstract the probability, the less certain the grounds for belief.

Condorcet specifically applied this to elections. The purpose of his *Essay* was to 'inquire by mere reasoning, what degree of confidence the judgment of assemblies deserves'. For instance, different systems of voting can produce different results with precisely the same electorate. A preferential system of voting (voters naming first and second choices amongst the candidates) will frequently produce a different result to a single choice (first-past-the-post) system. The latter gives a simple, clear-cut majority, but may not reflect the electorate's more varied sympathies. For instance, a 'Green' Party may be incapable of winning a seat in a single-choice election, despite there being a widespread groundswell of support for its aims. This would be reflected in a voting system which registered, and took account of, second choices. Condorcet also showed how a single-choice system could provide a result which nobody intended. He used complex tables to demonstrate his point; but instead we can use a simple example. Picture an electorate where the conservatives, liberals and communists all have one-third of the popular support. This could easily result in a liberal majority government, rather than the expected coalition. Many conservatives might vote liberal – anything to keep out the communists. And many communists might vote similarly to keep out the conservatives. Condorcet's analysis also applied to the jury problem. How do twelve people in fact arrive at a single verdict? What bargaining, personal assessment, reliance on expert opinion, etc., is involved? And 'what degree of confidence' does the jury's judgement 'deserve'?

Condorcet analysed how groups of people behaved when they made choices. To do this he employed much subtle and ingenious mathematics – with the obvious result. His work was largely ignored. Not until the present day has the significance of his findings become apparent for economics. Contemporary economists have come up with an unprecedented innovation. When constructing economic theories, they decide to take into account the people who play the central role in these theories. Who are these statistics, and how do they actually behave? Not for nothing has this branch of economics become known as 'impossibility theorem'. This is the work of the leading modern American economist Kenneth J. Arrow, who succeeded in proving what many may feel was obvious all along: namely, that human decision-making is not a rational process. Condorcet's understanding of 'subjective probability', with its differing 'grounds for belief' from 'abstract probability', was the first step towards analysing the complex reality of group decision-making which underlies all economics.

Despite his aristocratic lineage and his attendance at the Paris salons, Condorcet was a man of little social sophistication. Though presentable in appearance, his coolness of manner alienated many. Only when discussing ideas would he overcome his shyness and become animated. Only then, and when he fell in love – which seems to have happened on several occasions. Then the snow-covered volcano would erupt with overwhelming intensity. As a result, the object of his attentions usually fled. The women of the Parisian salons were used to men who intimated their desires in a more subtle and devious manner. Not until he was forty-two did true love at last triumph.

In 1786 Condorcet fell in love with and married the 22-year-old Sophie de Grouchy, who was widely regarded as the most beautiful woman in Paris. But this being a French love story, things were not quite as simple as that. To begin with, Sophie was already the mistress of the Marquis de Lafayette, the French hero of the American Revolution. Condorcet was forced to beg Lafayette for the hand of his mistress, but appears to have thought little of this humiliation.

Sophie was an intelligent and sympathetic personality. She too had a passionate interest in reform and economics. Soon after her marriage she embarked upon the first (and best) French translation of Adam Smith's *Theory of Moral Sentiments*. She also set about socializing her husband, and the salon she ran at their home soon became a major social and intellectual attraction: the favourite weekly meeting place of *les philosophes*. By now Condorcet had been appointed Inspector of the Mint, and Sophie's salons were held at the large town house which went with the job. This was the Hôtel des Monnaies ('Money Mansion'), an apt address for an economist.

Less than three decades earlier Louis XV's extravagant mistress

Madame de Pompadour had declared, *'Après nous le déluge'* ('After us, it will all be swept away'). From then on many in France had come to realize that the despotic *ancien régime* was on its last legs. But how long would it struggle on? And how would it end? The answer came on 14 July 1789 in Paris, when the mob stormed the notorious Bastille. Alas, inside they found only seven prisoners left to liberate – including an old roué charged with incest, and a lunatic who declared that he was God. Despite this setback, the French Revolution had begun.

Condorcet supported the Revolution, and was elected as a representative for Paris in the legislative assembly. Here, he presented a plan for setting up a national education system. He also expressed his other progressive views. In many of these he was far ahead of his time. Condorcet was anti-slavery, believed in votes for women, was antagonistic towards religion and was a staunch anti-monarchist. He continually reiterated his belief in 'progress and reason'. Such views are epitomized in a satirical article he wrote at the time. This suggested that France should replace the monarchy with an 'economical royal automaton'. This 'mechanical king' could perform ceremonial functions, and his behaviour could be adjusted to accord with changes in these functions. He could converse with visiting monarchs and could fulfil his other duties – all at a minimum of cost to the country (no extravagant dining, concerts, royal balls, etc.). He could automatically sanction all decrees passed by the legislature and automatically appoint ministers in line with the wishes of the assembly. Admittedly, the hereditary principle would be lost. But if this automatic monarch was well maintained, France would instead have an eternal monarchy. Previously the king had left himself open to ridicule when he declared himself infallible in his judgements and inviolable in his rights; now such superhuman powers could be claimed without absurdity. Condorcet's parable was in fact more than just satire. This was indeed a metaphorical blueprint for his idea of politics as science – all part of his belief in Progress.

In 1792 Condorcet drew up a proposal for a new French constitution. This was in line with the views of the moderate Girondists, but the Jacobins under Robespierre now took power and the Reign of Terror soon began. Condorcet was forced to flee for his life and went into hiding. He made use of this period of enforced inactivity to write his *Sketch for a Historical Picture of the Progress of the Human Mind*. In this, Condorcet maintained that human misfortune stemmed from ignorance (usually imposed by kings, priests or vested professional interests). The progress of reason led to an indefinitely expanding insight into the laws of nature. This would lead to increasing technological control over the forces of nature, the exposure of errors and the elimination of anti-social ideas. Human nature was not explicable purely by moral laws, or purely by the physical sciences. These two views would be merged to form a complete science of human nature. Here, Condorcet is laying the foundations of modern sociology, which plays such a vital role in present-day economics. It is also possible to recognize much of the modern world in these ideas. Our dreams (and our nightmares) are easily identifiable in Condorcet's all-embracing Progress.

Condorcet's economic thinking is embedded in his ideas on progress. He advocated increasing economic equality for all, and the granting of rights to protect workers from unscrupulous employers. More practically, he outlined an embryo welfare system which would provide for poor workers. He also sought government intervention to lower interest rates, thus enabling workers to borrow money without involving themselves in crippling costs. In other regards, his economic theories extend little beyond Physiocrat ideas. But his importance is in setting the scene. Progress was a new and powerful idea, a cause for optimism. Adam Smith had proposed his ideas with a static economy in mind. In practice the Industrial Revolution was rendering this notion redundant. Progress in Britain tended not to be regarded as an intellectual concept, as it was in France; yet its practical effects were plain to see. The Industrial Revolution was bringing about a social upheaval which gave little room for optimism amongst the majority of the people. The French Revolution brought about an even greater social upheaval – yet to begin with this gave rise to a new hope, both social and economic. The idea of economic progress inspired optimism in Condorcet. France believed in the mechanics of progress (a mechanical monarch and the like); Britain had machines in its factories.

Condorcet ended up taking clandestine refuge in a house on the southwestern outskirts of Paris, but soon become convinced that the place was being watched. Secretly he fled into the nearby woods of Meudon. After living rough for three days amongst the quarries, thickets and ponds he stumbled into Clamart at dusk on 27 March 1794. His ragged appearance quickly led to his arrest and imprisonment. Two days later he was found dead in his cell. The cause of his death remains uncertain. He may have been poisoned (or committed suicide with poison), but it is possible that he died of sheer exhaustion.

Across the Channel one of England's leading economic thinkers was leading the easy life of a young Cambridge don. Yet this life of comfort and privilege was to inspire a principle of the profoundest pesssimism, conjuring up visions far worse even than the excesses of the French Revolution and the Reign of Terror. To this day, the Malthusian Principle of Population remains perhaps the most chilling warning for humanity's future.

Robert Malthus was born in 1766, the son of a wealthy, eccentric country squire who lived near Guildford in the south of England. Squire Malthus had read Condorcet, and cherished woolly utopian ideas concerning the progress of humanity. He decided to educate his son himself, though later Robert was sent away to a like-minded tutor in the north of England (who would subsequently be imprisoned for supporting the French Revolution). Despite this wayward education, Malthus's intelligence ensured that he learnt enough to get into Cambridge, where he studied maths and natural philosophy (science). He excelled at these subjects, also establishing himself as a gregarious chap and a hearty sportsman. A handsome young man, he chose to wear his hair in long golden curls. The prevailing hairstyle of the period was the more austere pigtail, as sported by Lord Nelson, the hair lightly brushed with white powder to produce a distinguished greying effect. Malthus chose to powder his golden locks with rather more eclectic pink powder. From his eccentric father he had also inherited a cleft palate, which left him with slightly defective, high-pitched, nasal speech. After taking his degree he was elected a fellow of Jesus College, and then took holy orders. However, the Reverend Malthus was to become no mild benevolent parson, and he clashed heatedly with his father's philanthropic views. Their main point of contention was whether it was possible to alleviate the lot of the poor. The Reverend Malthus was utterly convinced that it was not, and embarked upon considerable research to prove his case. The result was a 50,000-word paper entitled 'An Essay on the Principle of Population, as It Affects the Future Improvement of Society, with Remarks on the Speculations of Mr Godwin, M Condorcet and Other Writers'. Father was so impressed by the brilliance of his son's essay that he had it published at his own expense – even though he continued to disagree with his son. Malthus's essay caused a sensation, and immediately established its author as a leading academic thinker.

Malthus was to rewrite his *Essay on Population* (as it is now commonly known) several times, but its central argument remained unchanged. Such great hopes for Progress as those put forward by Condorcet were doomed to disappointment. Why? Because the population would always increase faster than the means to sustain it. 'The power of population is so superior to the power of the earth to reproduce subsistence for man, that premature death must in some shape or other visit the human race.' Humanity was impelled by two irresistible forces: the need for food, and an insatiable sexual desire. As a result, the population was liable to increase in a geometric progression (i.e. 2, 4, 8, 16, 32 . . .) The means of subsistence, on the other hand, only increased in an arithmetic progression (i.e. 2, 4, 6, 8, 10 . . .).

Malthus was determined to replace what he saw as the cloud-cuckoo-land of his father's ideas with the harsh truths of reality. Economics must be empirical rather than speculative. It should aim to be a science, studying cause and effect with mathematical rigour. Malthus's principle concerning the geometric and arithmetic progressions of population and subsistence would appear to fit this criterion perfectly. Unfortunately, his use of statistics was somewhat less than scientific. One of the main works he drew upon to back up his argument was *Observations concerning the Increase of Mankind* by the great American pioneer scientist and diplomat Benjamin Franklin. In this, Franklin stated that the American population tended to double every twenty-five years, and in some isolated settlements this took just fifteen years. Malthus over-looked the fact that these included figures for immigration as well as home-born Americans.

The Industrial Revolution was meanwhile beginning to change the face of Britain. This was the period when Manchester was 'steam mill mad', and the backstreets of such mushrooming cities were beginning to fill to overflowing with a new urban poor. Prior to this period, the population had been regarded as contributing to the nation's wealth and strength. More people meant more workers, more soldiers. Now things appeared to be getting out of control, and the first doubts began to set in. What was happening? The fact is, no one really knew. The first census was not to take place in Great Britain until 1801. Prior to this the population could only be estimated. Local estimations, such as Graunt's seventeenth-century survey of London, were likely to have been fairly accurate. But when attempts were made to estimate the entire population, the figures varied widely. Prior to the 1801 census, estimates for the population of Great Britain (minus Ireland) varied between 9 and 12.5 million. (The census would reveal that the actual figure was 10.4 million.)

People were worried, and rightly so. If Malthus was right, the world was in for a population explosion, no less. Malthus may have been wrong in detail, but the drift of his argument was (and in part remains) inescapable. Population outruns the ability to support it. And if this is so, starvation would appear to be inevitable. But worse was to come. Any attempt to ameliorate the living standards of the poor, such as suggested by Condorcet, would only result in hastening the catastrophe. As the lot of the poor improved, they would inevitably produce more children, thus impoverishing themselves once more. Such social progress only caused an increase in overall misery. And the same was true of charity, according to this Christian pastor. It was Malthus's unrelenting pessimism which famously prompted the Scottish thinker Thomas Carlyle to dub economics 'the dismal science' (a detracting epithet which clings to this day, its users mostly oblivious of the particular dismal thinking to which it referred).

Waxing lyrical in his gloom, Malthus went on to claim that the population will always increase to the very limit of what it can sustain. Then 'positive checks' will come into play. These included all the old standbys of the misanthropes and disaster-merchants – famine, plague, war, disease, catastrophe and so forth. By now Malthus had taken off well beyond his empirical base and was arguing by the seat of his pants. Only the flimsiest of evidence was offered: truisms and historical conjecture: 'The effects of the dreadful Plague in London in 1666 were not perceptible 15 or 20 years afterwards.' 'The traces of the most destructive famines in China and Indostan are by all accounts very soon obliterated.' Yet despite this cavalier approach to his 'empirical science', Malthus had made an essential point, and he knew it.

So what was to be done? Here Malthus was ahead of his time. His remedies, his prescriptions for social and moral behaviour, his attitude towards the poor – all looked forward to the repressive mores of the Victorian era (which would not begin until 1837, four years after his death). In Malthus's view, the only way to avoid the population catastrophe was through sexual abstinence and delaying marriage. He even suggested that a proclamation should be read out at marriage cermonies warning couples that they would have to bear the financial burden and consequences of their passion. Even so, he remained doubtful of humanity's ability to curb its insatiable desire for sexual pleasure. The population was bound to increase. It was just not possible to bring about an overall improvement in the economic welfare of the nation.

Malthus's *Essay on Population* struck a nerve, and became an immediate topic of debate. Its effects were quickly visible. The social optimism generated by the Industrial Revolution amongst the middle and upper classes underwent a sobering re-estimation. Meanwhile, some employers saw Malthus's ideas as a justification for the subsistence theory of wages (despite Adam Smith pointing out its short-sightedness). Why pay workers more than the absolute minimum: this would only increase their misery. Likewise, it enabled many to avoid giving to charity – with a clear conscience. After all, this too only made matters worse. But Malthus's effect was not just amongst the disingenuous and the unscrupulous. The prime minister, William Pitt the Younger, had previously argued in favour of poor relief. The Poor Laws ensured that workers who earned below subsistence level received a dole. By the turn of the century almost one in seven of the population was receiving assistance of some kind, with £4.5 million being raised for this purpose by means of local property taxes. It was widely felt that the Poor Laws not only relieved suffering, but also encouraged larger families amongst the poor, thus adding to the national wealth. Two years after the appearance of *Essay on Population*, Pitt withdrew his support for poor relief, convinced by Malthus's arguments.

The Poor Laws were now seen as limiting the mobility of labour, by tying families to the parishes where they recieved their dole. And the encouragement to larger families was now viewed with horror. According to Malthus, had the Poor Laws 'never existed, though there might have been a few more instances of severe distress, the aggregate mass of happiness among the common people would have been much greater than it is at present'. Many felt that Malthus's argument had an incontrovertible logic. Others decided to dispense with such rationality. In the view of Robert Southey, who would later become poet laureate, Malthus was a 'mischievous booby' who could only take his place amongst 'the other voiders of menstrual pollution'. Such extreme reactions were understandable. Even at his most compassionate, the Reverend Malthus's words leave a nasty taste. As he saw it, the 'severe distress' of the unemployed should 'find some alieviation'. But this should be in workhouses, rather than 'comfortable asylums'. Here, they should be made to work, and the 'fare should be hard'. The aim should be to encourage such people to find work. But what if there was no work? What if these people were unemployed through no fault of their own – such as the widespread introduction of new machinery? These questions were not considered relevant to the overall scheme of things.

Fortunately, events have proved Malthus's 'incontrovertible logic' wrong – at least in Western Europe and North America, the places to which he most frequently applied it. Here, the increase in economic well-being has not resulted in a population explosion. Malthus needed economic stasis in order to avert a catastrophe. Yet by the early nineteenth century economic growth was even more inevitable than his 'incontrovertible logic' appeared to be. And with this growth came progress as well as widespread hardship. Malthus made no allowance for advances in medicine and contraception. As it happened, raising the standard of living would result in a decrease in the death rate, as much as an increase in the birth rate. The consequent increase in a healthy population was accommodated by an on-going agricultural revolution of unprecedented proportions. In 1700 agricultural productivity throughout

Europe had barely altered since ancient Greek times. During the ensuing century before the publication of *Essay on Population* it doubled. Malthus took no account of this.

On the other hand, Malthus's principle of population still has distressing relevance in the larger world beyond the confines of Western Europe and North America. Africa and other regions of the Third World remain no strangers to populations which have outstripped production. But increased agricultural production meant that famine had disappeared from Western Europe almost a century before Malthus came up with his 'harsh but necessary' ideas. (During the only major exception, the Irish Famine of 1845–9, Ireland still produced enough to feed itself. Famine was a consequence of Britain insisting that this vital supply was exported.) Yet the world population continues to expand at an alarming rate: 6 billion at the turn of the millennium; a projected 7 billion by 2010. Does this mean that Malthus's principle continues to apply on the global scale, and will soon be applying to Western Europe just surely as it did in the Middle Ages? Will technology be unable to keep up with the ever-growing demand for resources?

Thirty years ago leading US biologist Paul Ehrlich, author of *The Population Bomb*, agreed with Malthus. The end was nigh. Prices of energy and commodities would begin their inexorable rise as they became ever more scarce. In 1980 the US economist Julian Simon disagreed, and challenged Ehrlich to a now famous bet. Ehrlich could choose $1,000 worth of any world commodities, and wait for ten years. If these went up in price (indicating that they were becoming more scarce) Simon would pay him the same as the increase. If they went down, Ehrlich would have to pay Simon the same as the decrease. Ehrlich accepted the challenge, and purchased stocks of five metals worth a total of $1,000. By 1990 these were worth $424, and the optimistic Simon had won $576. Admittedly, there had been a large increase in world population. But this had been more than counterbalanced by other factors. Vast new oil fields had been discovered; there had been another agricultural revolution, allowing many Third World countries to feed themselves; and diminishing supplies of several metals were now being replaced by synthetic materials such as plastics.

Until recently India, China and Africa all experienced famine. Now this scourge is largely limited to Africa. Before 2010 it is possible that famine will have been eliminated here too. It is also likely that the global population will have begun to level out, just as it has in Western Europe and North America. Global resources are not unlimited, but we are some way from exhausting them. And despite much chronic urban overcrowding, far larger tracts of our planet remain uninhabited.

After preaching against the evils of marriage for many years, Malthus finally succumbed to the common fate of a humanity driven by its 'insatiable sexual desire'. At the age of thirty-seven he married the 27-year-old Harriet Eckersal. Many were shocked, imagining the creator of the Malthus principle of population to be a confirmed misogynist. Yet a contemporary acquaintance described him as 'a good natured man and, if there are no signs of approaching fertility, [he] is civil to every lady'. Regardless of this difficulty, Harriet Malthus managed to produce three children.

By now Malthus's *Essay* had made him famous throughout the land, and beyond. In 1805 he joined the teaching staff of the East India Company's college just outside London, where young administrators were trained for the Indian colonial bureaucracy. Malthus was appointed professor of history and political economy, making him the first academic economist named as such. It was during this period that Malthus met David Ricardo, the other leading British economist of the period. Though they disagreed profoundly on many economic matters, Ricardo and Malthus were to become close friends. In 1820 Malthus published *Principles of Political Economy Considered with a View to Their Practical Application*. The industrious Ricardo immediately made 220 pages of notes

132

correcting its economic errors, declaring that there was 'hardly a page [devoid of] some fallacy'. Despite this, Malthus had some pertinent things to say. Adam Smith had explained how profits came about, and how they fluctuated. Yet he had not come up with any general theory of profit. (Economics was still formulating itself, its theorizing only gradually penetrating the smoke of commercial reality.) Malthus justified profit by stating that this was the capitalist's return on his investment and risk, two essential features of his role in production. By investing his capital in tools and machinery, he improved the production of his workers – and thus deserved his profits.

However, Malthus's most important analysis was of 'general gluts', when the whole economy came to a halt. This was a daring topic to choose, for according to the economic orthodoxy of the period such a thing simply could not happen. In the view of the contemporary French economist Jean-Baptiste Say, the capitalist system could never break down, because it was essentially self-adjusting. He had formulated a law, still know as Say's Law, which was to remain orthodox thinking about markets for well over a century. Put simply, this stated that 'supply creates its own demand'. In other words, the value of all commodities produced will always be equal to the value of all commodities bought. Thus, there could never be a general overproduction – resulting in a 'general glut' of unbought goods. (In modern terms such a state of affairs is referred to as a depression or slump, when the market is swamped by too many goods, and there is no one to buy them.) There could, of course, be 'partial gluts'. But Say was of the opinion that any partial glut would soon be remedied by market forces – such as caused by the seller lowering his prices to clear the market of unsold goods. Malthus disagreed, pointing out that economic systems were subject to periodic 'gluts', which could in the end worsen into a 'general glut'. Prices rose when there was an increase in spending, and they also fell when there was insufficient spending. Yet low prices couldn't always be relied upon to clear the market, especially if consumers had insufficient money. In this case a general glut was bound to result. To remedy this problem, Malthus suggested that the state should then intervene in the distribution of income. Landowners should recieve more, and capitalists less. This would stop the capitalists overproducing. Landowners, on the other hand, would spend their excess income on more servants and luxury goods – thus relieving unemployment and creating demand. He also suggested 'the employment of the poor in roads and public works'. This would produce more income and get people spending again. In other words, the way out of a depression was by extra spending. This great (and uncharacteristically humane) insight would remain overlooked until well into the twentieth century, when it would be prescribed by Keynes as a cure for the Great Depression.

Malthus's treatment of gluts was radically ahead of its time in another aspect. Not only did he suggest a cure for them, but he had also indicated an important defect in the capitalist system itself. 'For the first time,' as the twentieth-century economist Eric Roll pointed out, 'the possibility of crises arising from causes inherent in the capitalist system was admitted.' Ricardo may have found Malthus's work littered with errors, but the greater error lay with Ricardo – who refused to admit such flaws, or the very existence of general gluts.

Just fourteen years after producing his second great work, Malthus died at the age of sixty-seven. Revered by some, demonized by others, the 'pitiless parson' remained a controversial figure to the end. In the words of modern American economist Todd G. Buchholz, 'When Malthus died, some came to the funeral to mourn, others to make sure he really was dead.' But of course he wasn't. In the very year of his death a new Poor Law was passed. This was in line with Malthus's economic thinking. It decreed that all able-bodied workers unable to provide for them-selves, and their families, could only receive relief if they surrendered themselves to the workhouse. Here, families were separated, conditions were made deliberately harsh, and all were were subjected to a regime of vindictive humiliation by sadistic staff. Inmates were frequently forced to live off slops and gnaw bones. Mothers and children were set to work picking oakum; and able-bodied men were made to crush bones for animal feed, alongside criminals and lunatics.

Although Malthus's immediate influence was baleful, his future influence would not prove quite so negative. In 1838 Charles Darwin read Malthus's *Essay on Population*: 'it at once struck me that under these circumstances favourable variations would tend to be preserved and unfavourable ones to be destroyed'. Malthus had accuratedly described the conditions in which the survival of the fittest could take place. As Darwin put it, 'for in this case there can be no artificial increase in food, and no prudential restraint from marriage'. What Malthus had in fact described was the animal kingdom before the advent of civilized humanity.

David Ricardo's view of society was, if anything, just as pessimistic as that of his friend and sparring-partner Malthus. Ricardo too viewed with concern the 'dark satanic mills' of the Industrial Revolution which were now spreading over England's 'green and pleasant land'. (During the half century of Ricardo's lifetime, the population of Manchester would triple to around 200,000.) Ricardo was in accord with Malthus's view of population, and it was he who formulated the Iron Law of Wages: any attempt to better the lot of the workers was futile, thus wages should remain at subsistence level. His profound analysis of society led him to equally gloomy conclusions. Society was not a large family, an all-embracing mechanism, or even a bitter struggle for survival – it was more a grim battle for supremacy between two powerful classes. The old established landowners were locked in conflict against the new rising capitalists, with the working class simply trampled underfoot.

Yet, as we shall see, Ricardo was not so overwhelmingly pessimistic as he would at first appear. His analysis of society and how it worked was the deepest since Adam Smith, and it was Ricardo who laid the foundations of economics as a systematic science. Smith had detected the drift of how economics worked: Ricardo saw below the surface. It was he who began to grasp the laws which governed this drift. Ricardo had a gift for intellectual abstraction of reality, such as had previously only been applied to physics. Or the Talmud.

And here lay the clue. Ricardo was part of the first great intellectual flowering of the Jews in Europe. The Jews were now beginning to take their part in mainstream society. Minds emancipated from the unending subtleties of Talmudic interpretation were now turning to the comparatively simple intellectual problems presented by social reality. This was the period which would produce the Rothschilds, Europe's premier banking family, as well as the philosopher Moses Mendelssohn, the poet Heinrich Heine, the composer Felix Mendelssohn, and Britain's first Jewish prime minister Benjamin Disraeli.

David Ricardo was born in London in 1772. His father was a Dutch Jew who had emigrated to Britain and begun trading on the London Stock Exchange. At that time only a dozen 'Jew brokers' were allowed to trade in the Exchange, and were restricted to a part of the floor know as 'Jews' Walk'. Despite these restrictions, Ricardo Senior soon made a fortune. His son David was brought up in a cultured Jewish household, but received no formal schooling. At the age of fourteen he went to work in his father's firm. Some are mathematical or musical prodigies, Ricardo was a financial one. His young, untutored mind was quickly able to grasp the most complex financial deals and strategies. He soon became the apple of his father's eye.

Then disaster struck. At the age of twenty-one David fell in love with a goy, became a Christian, and married this person (who turned out to be a pleasant, intelligent Quaker girl). With masterly

David Ricardo, whose financial expertise put
economics on a scientific footing and also made him
one of the richest men in Britian

understatement, it was recorded that 'Ricardo's mother was unable to accept this situation, and he also found it necessary to leave his father's firm.' But the penniless Ricardo was now free to set up on his own, no longer subject to the restriction on Jews. Using his contacts on the Exchange, he managed to raise £800, no mean sum in those days, when £1,000 would secure a country estate. Within five years he had become financially independent, and had sufficient wealth and leisure to educate himself. He could now afford to live like an English gentleman, and began reading literature, while his scientific interests led him to build his own laboratory and start a wide-ranging collection of minerals. Then, at the age of twenty-seven, finding himself bored one day during a holiday in Bath, he read by 'accident' Adam Smith's *The Wealth of Nations*. He knew at once that he had discovered his subject. For the next ten years, after his hard day on the Exchange he would pursue his hobby with increasing perceptiveness. Finally, in 1809 he wrote a letter to the *Morning Chronicle*, setting out his views on how to settle the current gold crisis. The originality and brilliance of his ideas were recognized at once, and he was soon contributing articles to the *Morning Chronicle* on a regular basis. A year later, these formed the basis of his first published work, *The High Price of Bullion, a Proof of the Depreciation of Banknotes*.

For seventeen years Britain had been waging war against Napoleon, in the attempt to contain the French domination of Europe. The financial strain of this war had begun to tell. In order to stem the drain on Britain's gold reserves, the government had decreed in 1797 that the Bank of England should no longer give gold in exchange for its banknotes. (British banknotes, such as the £10 note issued by the Bank of England, still bear the legend 'I promise to pay the bearer on demand the sum of £10'. This promise was first broken in 1797, and for much of the ensuing two centuries the Bank of England has continued to print lies on a larger scale than any tabloid.) Freed from the restraint of actually having to pay for their notes, the Bank reacted like anyone with a licence to print money. The excess of banknotes also meant that they could now lend more money. When prices began to rise, and the value of the pound fell, the Bank stoutly maintained that there was no connection between its actions and these subsequent events. Ricardo succinctly answered that there was. And he went even further, raising profound questions about the theory of banking. What precisely is a central bank? What is it for, and what should it do? In Ricardo's view, the Bank of England had a duty to conduct its policies in accord with prevailing economic conditions. The quantity of banknotes in circulation *did* affect prices. (More money chasing fewer goods.) And this had an effect on the foreign exchange rate – where payment *was* frequently required in gold. The bank's behaviour was thus responsible for a further drain on the country's precious gold reserves. The government agreed with Ricardo, and took action accordingly.

As a result of this work, Ricardo made three friends who were to play an influential role in his thinking. Firstly, he met the philosopher Jeremy Bentham, the founder of Utilitarianism. This moral philosophy was based on the premise that one should always act in such a way as to bring about the greatest happiness for the greatest number of people. Another man who was to play a leading role in Ricardo's life was the Scottish thinker James Mill (father of the other leading Utilitarian philosopher John Stuart Mill). James Mill had close connections with the leading radical politicians of the day, and was to be a source of political guidance for Ricardo. It was at this time that Ricardo also met Robert Malthus. The parson professor of economics and the successful young financier at once recognized each other's superior talents. No other contemporary thinkers had such a grasp of economic matters. However, the approach of these two supreme practitioners to their subject was diametrically opposed. Malthus based his ideas on experience (or liked to think he did). Ricardo's forte on the other hand was for abstract thought: the ability to construct theories – which, as we shall see, could fly in the face of common sense or apparent experience. The two quickly became close friends, visiting each other and corresponding at length on a regular basis. Their correspondence is a rare blend of amicable conflict and open closeness. As Malthus put it, 'I should not like you more than I do if you agreed in opinion with me.'

Their friendship contained many ironies. The Continental Jew had become a respected figure in English society, while the English parson was the object of controversy and prejudice throughout the land. And it was the theoretically minded Ricardo who gave practical aid to the empirical Malthus. An example will suffice.

When purchasing stock, Ricardo would often put some of this in his friend Malthus's name. If he later sold the stock at a profit, he would send Malthus his 'share'. Where economics was concerned, Malthus was the harsh pessimist. Where finance was concerned, it was Ricardo who was the hard man, while Malthus was all timidity. In 1815, on the eve of the Battle of Waterloo, Malthus panicked and wrote to Ricardo, asking his friend to sell any stock he had purchased in his name. Malthus feared that Napoleon would win, and the stock market would plunge. Ricardo did as instructed, selling Malthus's stock at a small profit. But Ricardo held on to his own stock, and after Wellington's victory his profits more than doubled.

This story is linked to one of the great legends of finance. One of Ricardo's Jewish colleagues at the Stock Exchange was Nathan Rothschild. By 1815 the international intelligence network of the Rothschild family was so efficient that Nathan heard of the victory at Waterloo several hours before Wellington's official envoy reached London. Rothschild entered the Exchange with a grim expression, and immediately began selling stocks. His astutely observant colleagues quickly surmised that Waterloo was lost, and prices tumbled as a wave of selling swept the market. Surreptitiously Rothschild had his agents snap up these bargains. A short time later news of Wellington's victory was announced, and prices soared. A considerable part of the early Rothschild fortune is said to have resulted from this ploy.

The year 1815 also saw a crisis over the Corn Laws, which governed the import of grain into Britain. During the Napoleonic Wars Britain had been blockaded by the French, causing the price of grain to spiral. Following the defeat of Napoleon, imports of cheap grain from the Continent caused the price of grain to fall. This led parliament to raise the tariff on imported grain, which caused widespread public outrage. Like Adam Smith, Ricardo was a firm believer in free trade, which led him to publish his *Essay on the Influence of a Low Price of Corn on the Profits of Stock*. In this masterly piece of analysis Ricardo laid bare what was actually happening. Raising the tariff on grain benefited the landowners, who obtained a better price for their produce. The price of bread went up, so workers had to be paid higher wages. This had the effect of decreasing the profits of the capitalist manufacturers, who had to pay their wages. The capitalists thus had less to re-invest in their companies, which meant that industrial growth suffered. In Ricardo's words, 'The interests of the landlords is always opposed to the interest of every other class in the community.' However, because the old landowners dominated parliament, and the new capitalists were under-represented, the Corn Laws would not be repealed for over thirty years. Ricardo's ability to see to the economic heart of the matter won him many friends, and even the respect of his enemies.

By now Ricardo had retired. He was only just over forty, yet he had accumulated a fortune of around £1 million – placing him amongst the hundred richest people in the country. Ironically, considering his contempt for landlords, he had also acquired considerable landholdings. The largest of these was Gatcombe Park in Gloucestershire, which he made his home. (This is now the residence of Princess Anne, the Queen's daughter.) Encouraged by his political mentor Mill, Ricardo entered parliament in 1819. In the manner of the period, he simply purchased the seat of Portarlington, a small town in the heart of rural Ireland. (He would not even have been able to do this if he had not converted to Christianity. Practising Jews would remain excluded from the British parliament until 1858 – yet within ten years another convert, Benjamin Disraeli, would be prime minister.)

The new MP for Portarlington did not speak regularly in the House of Commons. He appears to have suffered from a slight speech defect, which rendered his voice unusually high-pitched – much like his friend Malthus in fact, a resemblance which could well have played an unremarked role in cementing their friendship. Despite his attested charm in social circumstances, Ricardo seems to have possessed a rather retiring personality. Although his superior expertise was recognized throughout the House, even by the landowners whom he so depreciated, he would only proffer his advice when coerced. 'I have no hope of conquering the alarm with which I am assailed the moment I hear the sound of my own voice.' If only all 'experts' were assailed with such modesty.

By now James Mill had also managed to persuade Ricardo to commit the full range of his economic ideas to paper. These appeared in *Principles of Political Economy and Taxation* (1817). Ricardo's most important and lasting contribution was perhaps the law of comparative advantage. Almost all economic laws are basically elaborations of common sense; Ricardo's law is a notable exception. It originated from Adam Smith's insistence on the efficiency of the division of labour, but pushes this to its counter-intuitive limit. Even so, the crux of Ricardo's law can be illustrated by a simple example. Two shipwrecked sailors inhabit an island. Popeye is good at growing spinach, and at distilling his own hootch. Barnacle Bill is irritatingly hopeless at both tasks. Economic common sense tells Popeye that he would be better off living at one end of the island, and letting Barnacle Bill get on with his own haphazard existence at the other end. But Ricardo showed that he would be wrong.

For the purposes of our argument let us assume the equivalence in value, for this particular nautical economy, of one barrel of hootch and one barrel of spinach. Dividing his time equally between both tasks, Popeye can produce in one week 2 barrels of hootch and 4 barrels of spinach. Likewise employed, Barnacle Bill can only produce half this amount of both items. Thus, if Popeye lives in splendid isolation, he ends the week with a total of 6 barrels of produce, whereas Barnacle Bill ends up with just 3 barrels of produce. So the whole week's production for the entire island is 9 barrels. But suppose Popeye concentrates on what he is best at, and leaves Barnacle Bill to get on with what he is least worst at. In other words, each specializes in the activity where he has comparative advantage (or comparative less disadvantage). The weekly figures now come out as follows. Popeye produces 8 barrels (concentrating entirely on hootch), and Barnacle Bill produces 4 barrels (of spinach.) Total island productivity now rises from 9 to 12 barrels.

Extend this simple model to an international scale: instead of Popeye and Barnacle Bill we have two countries. At once it becomes clear that isolationism is essentially inefficient, and free trade results in more production. Ricardo's law of comparative advantage is also true in the more complex conditions of the actual world economy. That is, it works for situations involving more than two countries, and also for trade involving more than two products. A country is always best off specializing in whatever field it has a comparative advantage (or least disadvantage). In the real world this results in more anti-commonsense situations. For instance, a country is not always best off producing what it does best (this is not necessarily what it does least worst). Britain's economic forte may have been producing ships, but the Japanese did this better. Britain was thus well advised to divert its energies from its greatest talent. Instead, it was best off concentrating on activities where it had a comparative advantage over the rest of the world, such as the manufacture of supersonic aeroplane engines and making marmalade. Japan is now learning this lesson too. As South Korean shipbuilding becomes more efficient, the Japanese switch to entrepreneurial international investment. Directly, or indirectly, they now finance the very yards that are putting their shipbuilding industry out of business. And according to Ricardo, they are right to do so. They may still be more experienced than the Koreans at shipbuilding, but their comparative advantage over Korea lies in finance. (This of course refers only to the so-called *economic* equation. The human suffering involved is not quantified. What part this should play in the economic equation remains of course a matter of heated debate – or whatever a democratically elected government thinks it can get away with.)

Over a century and a half after Ricardo first formulated his law of comparative advantage, this remains the basis of international trade. Indeed, many regard it as one of the essential insights into how economics works. It is easy to see how sophisticated elaborations of Ricardo's law, by modern-day wizards, now provide the springboard for much international and multinational economic strategy. And despite the fact that Ricardo's law has a *categorically* different view of economics, it is possible to discern for the first time a hint of the minimax approach that would one day characterize von Neumann's application of game theory to this subject. Go for the option that gives you the minimum maximum loss.

Ricardo also extended the ideas he had first put forward in his pamphlet on the Corn Laws. His laudable ambition was to discover some principle which might indicate how the 'social product' should be divided amongst the 'three classes of the community'. These were the landlords (who received rent), the workers (who were given wages) and the owners of capital (who gained profit). Here was the conflict which lay at the heart of society. As we have seen, if the landlords gained, it was at the expense of the other two classes – and thus society as a whole lost out. If the capitalists gained greater profits, they could invest these in expanding their business and providing more jobs for workers. This benefited two classes – and thus society as a whole benefited. If on the other hand the workers were paid higher wages, this lowered the capitalists' profits. Also (according to Malthus's principle, which Ricardo believed in) it led to an increase in population, a rise in food prices, more widespread misery, and so forth. For this reason Ricardo formulated his Iron Law of Wages, which decreed that workers should be paid subsistence level and no more. This was for their own good, as well as for the good of society as a whole. However, this is neither so repellent, nor so pessimistic, as it might at first appear. Adam Smith had tempered his subsistence theory of wages, by showing how it was more efficient for wages to be increased to the level where a worker could raise and maintain a family. Ricardo went one step further, insisting that the meaning of subsistence 'essentially depends upon the habits and customs of the people'. As living standards have increased, so has the acceptable subsistence level. Present-day welfare is paid at what modern society regards as subsistence level. Even a lifetime ago, such things as electricity, indoor baths and plumbing, TV ownership, and even the ability to maintain an inexpensive car, would all have been considered luxuries. Nowadays, in the modern Western world, such things are regarded as essential to subsistence. Ricardo's caveat regarding custom at least tempered his Iron Law.

However, it did undermine its use as an unchanging basic principle. This Ricardo sought to establish in his theory of value. According to this, 'commodities derive their exchange value from two sources: from their scarcity and from the quantity of labour required to obtain them'. Ricardo surprisingly belittled the former category, maintaining that scarcity only applied to goods which cannot be manufactured or reproduced – such as ancient coins, oil paintings and the like. These were irrelevant to the main production of society, which meant that labour was the central determinant. This essentially just notion – the labour theory of value – has a long philosophical pedigree. Indeed, it is said to have originated in the thirteenth century with St Thomas Aquinas. But 500 years had passed since the comparative economic simplicity of the medieval era. Ricardo was forced to take into account the on-going transformation being wrought by the Industrial Revolution. Adam Smith's pin factory had now given place to cavernous textile mills lined with rows of machines (such as the latest looms, operated by thirteen-year-old girls, which could do the work of twenty skilled weavers). Ricardo distinguished between direct labour and indirect labour. The latter involved machines. Since these machines themselves were manufactured, their value too could be calculated in terms of the direct and indirect labour needed to manufacture them. By further analysing this indirect labour, one could eventually calculate the amount of direct labour required to produce all manufactured goods.

Yet as already hinted, this theory of value had a basic flaw from its very inception. Even in the essentially static economy of the Middle Ages it would have been undermined by the supply and demand of the market-place. In other words, the labour theory of value may be how we would *like* things to be – but they aren't. As with so many general principles (not only in economics), the closer one examines them, the more the sand spills between one's fingers. A market economy simply doesn't work without supply and demand. Ricardo was, of course, well aware of this problem, but dimissed it as 'comparatively of very slight effect'. Initially, he admitted that the labour theory of value 'is not rigidly true, but is the nearest approximation to truth . . . of any I have ever heard'. Later he became convinced that his whole new scientific economics needed to be based on some truth which *was* rigid. To the end of his days, as the last grains of sand spilled between his fingers, he sought vainly to establish his theory as the 'absolute value' upon which economics could be based.

The end came unexpectedly and swiftly. In 1823 Ricardo retired from parliament on account of ill health. He returned to his beloved Gatcombe Park by stagecoach (George Stephenson would not open the first railway for another two years). By the end of the year he was dead, at the age of fifty-one. In his will he left money to his old friend and rival Malthus, who wrote, 'I never loved anyone out of my own family so much.'

Thus died the finest economic theorist since Adam Smith. Ricardo was the man who focussed Smith's ideas to the point of theoretical clarity. Insofar as economics can claim to be a science, it was founded by Ricardo.

6

Brave New Worlds

In 1815, during the height of the Corn Law debate, one member of parliament had declared with perceptive cynicism, 'The labourer has no interest in this question; whether the price be 84 shillings or 105 shillings a quarter, he will get dry bread in the one case and dry bread in the other.' This was Alexander Baring, son of the founder of the exclusive Barings Bank, the only real rival to Rothschilds in the City. (Seventy-five years later Barings would be begging the Bank of England to pay off its colossal debts of £21 million. The bank duly obliged, and Barings continued on its illustrious way. A century later financial incompetence, and the activities of the rogue trader Nick Leeson, resulted in Barings facing what was referred to this time as a 'black hole' of astronomical debt. In a memo against any further waste of public money, one of the Bank of England's advisors is said to have returned Alexander Baring's hoary sentiments: 'The Bank of England has no interest in this question; whether Barings' debts be 21 million or 21 billion, it will get dry bread in the one case and dry bread in the other.' Barings went under: comeuppance had been a long time coming.)

Back in the early nineteenth century, however, there were many who were not prepared to wait so long for justice. Even if the poor were forced to eat dry bread, they surely had an interest in how the economy was run. Economics was developing largely of its own accord. Among others, Smith and Ricardo had come up with perceptive ideas on how it worked, and how this could be improved upon. The invisible hand, free trade, comparative advantage, and so forth. But there was no ultimate reason why economics should develop in this fashion. The Industrial Revolution had brought about widespread suffering, and disproportionate benefits for the few. In the mines semi-naked women and children dragged sledges of coal, sloshing in the fetid dark on all fours along narrow shafts far underground. Meanwhile the Prince Regent entertained Mrs Fitzherbert and Beau Brummel to fifteen-course dinners amidst the oriental exotica of the Royal Pavilion he had built by the sea at Brighton. Such heaven and hell circumscribed the harshness and quaint sentimentalities of an early Dickensian world. Surely there had to be a better way of doing things? A number of economic thinkers now began to come up with a variety of alternatives, ranging from the practical to the plain potty.

As a person Saint-Simon comes perilously close to the latter category. His economic ideas, on the other hand, were both original and humane. At least for the most part. Saint-Simon's unsystematic method left his ideas open to a wide variety of interpretation. He has, with some justice, been claimed as the rationalist thinker who became the father of socialism, and thus socialistic economics, as well as one of the founders of sociology and philosophical positivism. With equal justice he has been called a mystic, a totalitarian and a capitalist. So what exactly was he? In the most general sense, it could be said that Saint-Simon tried to combine the Christian view of life with the modern scientific vision, so that humanity could progress into the new industrialized future in a humane fashion.

Claude-Henri de Rouvroy, Comte (Count) de Saint-Simon was born in 1760 into the impecunious tail-end of one of France's most noble families. The Rouvroys insisted that they were direct descendants of Charlemagne – a bogus claim which Saint-Simon would be happy to promulgate. However, there was no denying that he was distantly related to his famous namesake, the eighteenth-century Duc de Saint-Simon, whose posthumous memoirs had dished the dirt on court life at Versailles during the era of the Sun King, Louis XIV. In the words of one commentator, Saint-Simon was born of the 'best but also most degenerate blood of France'. The nineteenth-century literary historian Émile Faguet would feature Saint-Simon heavily in his masterwork *Le Culte de l'incompétence*, characterizing him as '*un fou très intelligent*' ('a very intelligent lunatic'). This would seem to be harsh, but then who are we to disagree with someone who actually saw him at work?

The beginning was sane enough: education by private tutors, followed by entry into the army at seventeen. In 1781 he crossed the Atlantic with Lafayette's volunteers to help the Americans defend their newly won independence against the British. By now the young count had risen to the rank of captain. With 166 artillerymen under his command he played a vital role in the successful siege of Yorktown, whose surrender marked the bitter end for the British. As a result he received a personal letter of thanks from George Washington and was awarded the Order of Cincinnatus, the last Frenchman to receive this honour. Saint-Simon was particularly struck by the American army, which functioned efficiently and enthusiastically without having aristocratic officers – though it would be some time before he began to understand the full social implications of this.

Saint-Simon now left for Mexico, where he attempted to interest the viceroy in an ambitious scheme to build a canal linking the Atlantic to the Pacific. When this was rejected, he took ship for Spain, where he had more success with a scheme for a canal between Madrid and Seville, thus linking the capital to the sea via the river Guadalquivir. Six thousand men were earmarked for the task of digging the 250-mile canal across ranges with mountains rising to well over 4,000 feet. It was then discovered that Saint-Simon's 'extensive plans' consisted of little more than the idea itself. This meant that he was back in France in time for the Revolution in 1789. The aristocratic Comte de Saint-Simon immediately transformed himself into 'Citoyen Bonhomme'

(Citizen Goodfellow). In order to encourage peasants to take on their own land, the revolutionary authorities began putting up for sale land seized from the Church and aristocratic landowners. Borrowing money from a friend, Citoyen Bonhomme began snapping up large tracts at bargain prices. His local town then elected him to the National Assembly in Paris, where he put in a bid for the roof of Notre Dame, with the aim of stripping it of its vast quantity of lead. This (and one suspects a few other things) eventually led to his arrest when Robespierre launched his Reign of Terror. During the year of the Terror 300,000 were arrested and nearly a tenth of these ended up on the guillotine. Saint-Simon spent almost a year in conditions of grotesque privation, under daily threat of death. Then the Terror was over, and the prisons were flung open; it was the turn of the fallen Robespierre and his cronies to take up residence in the dungeons awaiting the guillotine. The Revolution had entered a new stage: the way was now open for Napoleon.

Saint-Simon emerged from prison to discover that he was a very wealthy man. Land prices had soared after the collapse of the Revolutionary currency. (The authorities had made the mistake of resurrecting some of John Law's ideas.) During his time in prison, Saint-Simon had had a dream. In the course of this, Charlemagne had appeared and addressed him: 'My son, your success as a philosopher will equal those which I achieved as a warrior and a statesman.'

Saint-Simon decided to invest his fortune in fulfilling this worthy project. He bought a mansion in Paris close to the prestigious École Polytechnique, so that he could attend its public lectures. He was now thirty-five years old and wished to make up for lost time. His declared intention was to know everything it was possible to learn: then he would launch himself as the Charlemagne of philosophers. At night he invited the leading philosophers, mathematicians and politicans to dine at his table on the finest cuisine and vintage wines. Politicians and scientists were encouraged to remain as guests in his house, in order to tutor him. Having spent several years of carousing and dissipation in army barracks in his old military days, Saint-Simon had acquired a taste for enjoying himself. He saw his mansion as a centre of high living and high talking. (Others, with some cause, soon came to regard it as a centre of extravagant debauchery.) A portrait of Saint-Simon from this period depicts a fashionably dressed round-faced man with a rather prominent nose, but an otherwise bland, strangely anonymous appearance. The man had yet to become himself.

Having experienced life at the extremes – on the battlefield, in the National Assembly, in the dungeons, at home – Saint-Simon eventually decided it was time he experienced how other people lived. So he got married. However, he took the precaution of drawing up a three-year contract for this arrangement. After only twelve months he had decided that his wife was nothing but an idle chatterbox (he couldn't get a word in edgeways), and that her guests were little more than ignorant roisterers (as distinct from his more intellectual cronies). The experiment of living a life of normalcy was abandoned, and in 1802 he set off for Geneva. His intention was to marry Madame de Staël, who was widely regarded as one of the finest minds in Europe. Her writing was influenced by a heady mixture of Rousseau's enthusiastic romanticism and Montesquieu's incisive rationalism – as indicated in the title of her major work, *On the Influence of the Passions upon the Happiness of Individuals and Nations*. In the words of the historian Robert Escarpit, she 'helped the dawning nineteenth century to take stock of itself'. Madame de Staël's salon attracted some of the greatest thinkers in Europe, and it now attracted Saint-Simon. He too wished to play his part in the new century. (And amazingly, he would: but this still lay far in the future.) Having presented himself to his hostess, Saint-Simon blithely informed her: 'Madame, you are the most extraordinary woman in the world, and I am the most extraordinary man. The two of us could have a still more extraordinary child.' But the prospect did not appeal, and Saint-

Simon was forced to abandon his project for extraordinary family life.

Yet it was this farcical visit to Geneva which produced Saint-Simon's first major work, *Letter from an Inhabitant of Geneva to his Contemporaries*. The originality and impracticality of his ideas is immediately apparent. Saint-Simon called for a complete reorganization of society. Although the nobility in France had been displaced, a new land-owning class had now taken their place. This stratum of society too would have to be replaced, as it was an impediment to social progress. 'Men must work,' declared Saint-Simon (who had not actually spent much time at this occupation himself). Society could lose 30,000 landowners, judges, ministers of state and the like with impunity, as these were just ornaments. But if it lost a mere 3,000 of its top functionaries, this would be a catastrophe. Society needed its physicists, its chemists, its mathematicians, its physiologists and its engineers. Indeed, scientists should take the place of priests in a new 'Religion of Newton'. For all its originality, Saint-Simon's thinking is very much a reflection of its time. The Revolution had taken place in France; many felt that it would now spread throughout Europe. Here lay a golden opportunity. Civilization could be completely reformed. Everything was in the melting pot: it would be possible to start from scratch with a rationally devised society. The theory of political economy could be completely rewritten to produce a more just distribution of the nation's wealth.

Saint-Simon sent a copy of his *Letter* to Napoleon, but it was ignored. The earlier Napoleon, seen throughout Europe as a liberator of the people, could possibly have been open to such ideas. But this was 1803. Napoleon had already signed a concordat with Pope Pius VI, reconciling the Revolution and the Church, and within a year he would declare himself Emperor (prompting Beethoven to tear the dedication to Napoleon from his Eroica (Heroic) Symphony: a typical reaction amongst Europe's disillusioned intellectuals).

Within two years Saint-Simon had squandered his entire fortune. The man who would change the world was now reduced to penury – prompting his enemy De Lepiné to remark, 'Regard the Count, descendant of Charlemagne ... reduced to begging in the Tuileries, bowing his head as he receives a sou in his hat.' Saint-Simon managed to find employment as a humble clerk in a government pawnshop at Mont-de-Piété. (The entrance to the building still has a plaque recording this ignominy suffered by the 'most extraordinary mind in the world'.) Fortunately, Saint-Simon ran into his old manservant Diard, who was so overcome by his former master's plight that he put him up in his modest lodgings. Whereupon Saint-Simon set to work once more, bombarding the world with pamphlets explaining his ideas. Their publication was subsidized by friends; and they were sent out to further friends and potential patrons, accompanied by notes in his own hand: 'I am dying of hunger . . .' 'I have subsisted for the last fortnight on bread and water . . .' 'All my clothes have been sold to pay for this work . . .' Only gradually did he manage to obtain subsistence-level stability.

Saint-Simon's ideas are unsystematic, and often barely coherent. But their hope and amazing prescience shines through. Many of his wild ideas, shorn of their wildness, are now woven into the fabric of our social understanding. He was determined that the harsh scientific principles of economics discovered by Smith and Ricardo should be tempered by the no less scientific principles of rational humanism. Enthusiastically he embraced the idea of progress which had been proposed a few years previously by Condorcet. The way Saint-Simon saw it, the world was now moving forwards into a new industrial age, where social problems would be solved by science and technology. This would be achieved by the scientific study of society. Only science and the application of scientific method could bring about this utopia, whose very essence would be economic.

In this ideal world, society would operate like a giant factory.

Everyone would have their own particular job to do, and those in charge would organize the more productive running of the factory. The government would be concerned with economics, rather than politics. Its concern would be efficiency, rather than power over the people.

It is not difficult to see echoes of this in present-day society. And indeed, the *material* conditions which prevail in the advanced industrial countries of today would have appeared to Saint-Simon as little more than an extension of his wildest utopian dreams. His critics simply jeered at such outlandish fantasies. Unfortunately, like Saint-Simon's vision of the Panama Canal, the particulars of how his ideas were to be achieved are largely missing. This vagueness has left his ideas open to wide interpretation. It is possible to see his vision of a scientific world as a totalitarian nightmare. It is also possible to see it as the dream of socialism. Indeed, in the twentieth century the dream and the nightmare would become one in Soviet communism.

It was Freud who said, in his beleaguered early career, that later generations would read his work and wonder why he was famous: all his original ideas would have become accepted as obvious. Many of Saint-Simon's ideas have suffered a similar fate. It is the sheer originality of his ideas which we now fail to recognize. While others looked on in dread, he welcomed the coming industrial age – foreseeing that it heralded a golden era for humanity. (This 'golden era', which we now inhabit, may be a very long way from being perfect, yet there is no doubting that it *is* a golden era. As that perceptive social critic P. J. O'Rourke recommended: those who consider any previous age was better to live in than this one should first contemplate the word 'dentistry'.) The 'religion of science', 'belief in technology', 'worship of progress' – such notions all originate with Saint-Simon. Likewise, he was perhaps the first to recognize the major role which economics would play in history. No one previously believed that economic progress would literally transform the world – as it had never been transformed before.

Again, the details are missing. But it is fair to say that what Saint-Simon foresaw was no egalitarian utopia – more a hierarchical socialistic society. According to Saint-Simon, the driving force of this society would be a class of entrepreneurial scientists. With hindsight it is easy to see that he is here confusing two separate roles. The scientist discovers, creates the technology: his role is in production. The entrepreneur enables this production to take place, and ensures that its product reaches the market place. Saint-Simon's originality is in his recognition that science and commercial enterprise would be the twin spearheads in a future industrialized society.

Saint-Simon secured his financial survival by editing a series of journals, most of which became defunct after a year or two. These were largely concerned with political, economic and philosophical ideas. Saint-Simon was assisted in the writing of these journals by a tragic and impecunious young student. This was Auguste Comte, who would later be recognized as the finest French philosopher of his time. Yet Comte's life would remain tragic and impecunious to the end: marriage to a prostitute, impotence, a bout of insanity, public ridicule of his life and ideas, attempted suicide – such was the parodic existence of nineteenth-century Parisian genius. (His contemporaries de Nerval, Baudelaire and Verlaine all lived remarkably similar lives, though, as these were poets, prodigious amounts of absinthe and opium were also involved. Despite Comte's 'profound agitations', to say nothing of the occasional 'cerebral explosion', he appears to have taken alcohol and narcotics in untypical moderation.)

Saint-Simon was already fifty-seven by the time he met the nineteen-year-old Comte in 1817, but this was to prove a major event in the lives of both these remarkable thinkers. They quickly discovered that they shared an enthusiasm for Condorcet, and their rapport was such that they were soon writing articles in each other's name, developing their ideas in tandem. Saint-Simon was certainly the mentor, but there's no doubt that Comte contributed significantly to this dialectical development. Saint-Simon's ideas evolved and clarified. The notion of sociology crystalized: 'the science of social organization' was seen as 'a positive science'. The latter phrase was in fact coined by Madame de Staël, but in Saint-Simon's hands it became an active principle involving laws derived strictly from observation and experience. Only such positive facts would be acceptable to this new philosophy, which Saint-Simon christened 'positivism'.

Saint-Simon foresaw a future when, by the application of mathematics to this positive science, humanity would be able to predict 'all successive changes'. The previous history of the sciences had indicated that such advances in society's predictive abilities were inevitably linked with the 'progress of the human mind'. In ancient times, humanity had learnt how to predict the movement of the stars, but astronomy had only begun to make proper scientific use of this knowledge when it had discarded astrology. Likewise chemistry, whose reactions were more difficult to predict, had only advanced when it had shed its belief in alchemy – another step in the progress of the human mind. Even more complex was physiology, which still had to divest itself of certain beliefs imposed by 'philosophers, moralists and metaphysicians'. But there was no doubting that the time would come when 'all successive changes' would be predictable. The deterministic workings of both science and economics would be revealed. This new scientific society would be run by its most enlightened members. These 'men of genius' would of course be scientists, the finest of whom would be mathematicians.

Before we dismiss Saint-Simon's elitism out of hand, it's worth remembering his historical situation. During the Revolution France had lost its entire ruling class, as well as thinkers ranging from Condorcet to Lavoisier (the Newton of chemistry). A potent political force had been the mob – which had given rise to the likes of Robespierre (whom Saint-Simon had good reason to detest) and then Napoleon (whose increasing megalomania and militarism

Saint-Simon had come to abhor). Not surprisingly, the mob inspired little sympathy in Saint-Simon, who had not detected much scientific thinking or mathematical activity amongst this bloodthirsty, bewhiskered rabble and its screeching harridans. The way Saint-Simon saw it, science would civilize us – and to a large extent he is right. That rabble and its harridans have become us. Science has not been without its catastrophes, yet its application to civil technology and economics has had the greater effect. It is this which has civilized us. We behave in a civilized fashion because we have more to lose by not doing so, rather than because we are morally superior human beings. Here, we approach the core of economics, and as we do so we see the unmistakable element of game theory emerging. We choose the minimum maximum loss. Civilization as a minimax phenomenon!

The further development of Saint-Simon's philosophical ideas was to be undertaken by Comte, who is now generally regarded as the founder of positivism. This involved the rejection of religion, metaphysics and superstition as outdated, pre-scientific ways of thinking. Instead, humanity should adopt a positive scientific approach to the world, involving only rational and empirical modes of thought. Despite attracting such devoted friends, Saint-Simon's ideas remained largely ignored. This hurt him deeply. He was an enthusiast: 'Remember that in order to do anything one must be impassioned.' Yet he retained an unsentimental streak: 'A man of genius who is beaten and cannot hope to be victorious in a new battle should kill himself.' As he entered his sixties he became increasingly disillusioned, and in 1823 he attempted to shoot himself. According to all reports, he made a grotesque bungle of it. Six bullets lodged in his brain, yet all he put out was an eye. He lived on, in poor health and all but destitute. His writings now began to emphasize the spiritual element that had always been latent in his ideas. Perhaps inevitably, there was a break with Comte. In 1825 Saint-Simon wrote *The New Christianity*, in which he sketched out his dream of a quasi-Christian socialistic society.

Socialism and Christianity were essentially the same: their ulti-mate concern was to support and uphold the downtrodden and the poor. Saint-Simon died in the same year at the age of sixty-four – an abject failure, his ideas ignored by all but his few faithful disciples.

It was only in his after-life that Saint-Simon achieved the exalted status for which he had always longed. The form this took was aptly idiosyncratic. His disciples began to spread the word about the new 'industrial religion'. This fulfilled a need amidst the spiritual emptiness of the new factory age. Saint-Simonian communities were set up, in which there was no private property; two bankers amongst the Saint-Simonians took over *Le Globe*, an influential newspaper of the period. Those who emphasized the spiritual aspect of the master's teaching founded the religion of Saint-Simonism, which soon established groups of churches throughout France, in Germany and in England. In Germany, Marx and Engels saw the embryo of their future ideas. In England, John Stuart Mill and Thomas Carlyle were similarly struck. In France, even Napoleon III became an admirer.

Saint-Simonian ideas were often open to contradictory interpretations – some of which had been espoused simultaneously by the master himself. The founder of a new religion was also the founder of French socialism. Some communities shared all property, including their women (and men, presumably). Others embraced chastity. Many adopted a uniform consisting of a blue tunic and red beret. According to the American economic historian Robert Heilbroner, some even 'wore a special waistcoat that could neither be put on nor taken off unassisted and that thus empha-sized the dependence of every man on his brothers'. Others of a more capitalistic bent saw a niche in the market, and proceeded to apply Saint-Simon's ideas to banking. In 1848 the Crédit Mobil-ier was set up. As its name implies, the aim of this bank was to provide quickly obtainable credit for new commercial enterprises. This became a huge success, helping finance the building of railways all over Europe. Shares increased dramatically; but as the bank was essentially a credit facility, it was always hampered by small resources. In 1865 the need for new funds became tantamount, and the bank decided upon a 'note issue' of long-term bonds, which was based upon much the same principle as fiduciary banknotes (i.e. the bonds were not backed by anything). The government quickly spotted the similarity to Law's banking methods and barred the issue. A year later the Crédit Mobilier went bust, leaving unfinished railways from Bulgaria to Switzerland.

Saint-Simon contributed many positive ideas to the history of economic thought, yet his lack of contribution to economic analysis has been seen as an almost equal minus quantity. Saint-Simon is typical of the figures thrown up between one age and the next (in this case, the *ancien régime* and the firmly established Republic). At such times of discontinuity everything seems possible. Smith and Ricardo had done little more than throw up their hands in horror when confronted with the injustices of the economic system. Saint-Simon and Comte had sought to remedy these. Their largely unsystematic belief in progress and positivism produced a wide variety of alternative ideas. They didn't succeed in turning the industrial inferno into a technological paradise, but their ideas gave birth to a growing belief that something could be done. For instance, the compassion of Saint-Simon's Christian-socialist idea would inform the founding of the welfare state. From now on, economics would involve an unavoidable moral strain.

Once again French imagination was matched by British pragmatism. Robert Owen, regarded by many as the founder of British socialism, was in many ways a complementary figure to Saint-Simon. Unlike Saint-Simon, he produced few original economic ideas – instead, his concern was with practical economic policy. As much as Saint-Simon's ideas found ready followers in France, Owen's practical efforts would pioneer a change of heart in Britain. His influence would gradually transform the bitter realities of raw economics into a somewhat more equitable process. Laisser-faire was all very well, but economics should not simply be left to benefit the few at the expense of the many.

Robert Owen was born in 1771 and grew up in the remote small town of Newtown amongst the mountains of mid Wales. His father was a local saddler, and young Robert was the sixth of his seven children. At school he learnt little: his master was a certain Mr Thickness, who appears to have lived up to his name. At the age of seven Owen was promoted to teacher status – whereupon he learnt nothing more in school. He left at ten, soon afterwards catching the coach to London to seek his fortune. According to his autobiography (written over seventy years later), he lived with his brother and took a variety of jobs. Here, Owen quickly grasped the ways of business, but was probably not quite such an exceptional businessman as his hindsight would have us believe. Yet he did possess two exceptional qualities, which were to stand him in good stead throughout his life. He read widely, and he had an ability to charm his seniors. According to his autobiography, by the age of ten his widespread reading had led him to conclude that there was 'something fundamentally wrong in all religions'. Despite this, he would retain a deeply compassionate – quasi-religious – regard for his fellow human beings. At the age of nineteen he managed to talk his way into the job of supervisor at a new cotton mill in Manchester. He made no secret of the fact that he had no previous experience of such work, but quickly remedied this by applying all his considerable energy and imagination to the job. Within a few years he was manager of one of the largest cotton mills in Manchester, where his industry and innovations soon led to him being taken on as a partner.

In the course of his work he made frequent visits to Glasgow, where he met David Dale, a leading Scottish banker and industrial-ist who also had interests in cotton. Dale had a cotton mill at New Lanark, a riverside village set amidst picturesque hills and woodlands twenty-five miles up the Clyde valley. Dale was something of a philanthropist, and would rescue orphans from the poorhouses of Glasgow to work in the countryside at his mill. The sight of New Lanark, and Dale's ideas, seem to have crystallized a growing dream of Owen's. 'Of all the places I have yet seen, I should prefer this in which to try an experiment I have long contemplated and have wished to have an opportunity to put into practice.' Owen set off back to Manchester, where he persuaded his partners to let him purchase the mills at New Lanark. Persuading the redoubtable Robert Dale to part with his mill was not such an easy task. Dale didn't like Englishmen. When Owen pointed out that he was in fact Welsh, this was dismissed as beside the point. He was a 'sassenach' (the Gaelic word for 'Saxon', used dismissively to refer to anyone from south of the border). Even more inauspiciously, Dale was also the leader of a very strict sect of Presbyterians, known as the 'Old Scotch Independents'. And here was this 27-year-old sassenach free-thinker who not only wanted to buy his mill, but also had the temerity to inform him that he'd fallen in love with his daughter. A drawing of Owen from this period depicts a fresh-faced young man with a protruberant nose, wily eyes and swarthy Welsh features: he was by all accounts 'no beauty'. So it speaks volumes for Owen's charm that he managed to win the heart of both father and daughter. In 1799, with a £3,000 dowry added to his £3,000 savings, Owen moved to New Lanark determined to make a go of both the business and his dream. He would transform this Scottish village into a community run on humane as well as productive lines: an economy that worked for all who lived in it.

New Lanark had 2,000 inhabitants, all of them dependent upon the mill. Five hundred of these were orphans from the Glasgow poorhouses, with children from six years up expected to work a thirteen-hour day. Dale's philanthropy had only been relative. Entire families lived in stone terraces of one-room cottages with no sanitation. The cobbled streets in between were slimy with rotting refuse. It was miles to the nearest shop, so all produce had to be bought at the company store – which set its own prices, thus recouping much of its outlay on wages. Such conditions quickly bred despair, and all the usual consequent social ills. Drunkenness, theft, brawling and other vices were rife. Working hours were often from before dawn till after dark, with no healthcare, or any provision for education. (In Dale's mitigation, it must be said that conditions elsewhere were mostly far worse.)

Owen immediately set about building a second storey to every house. He refurbished the store with standard (rather than sub-standard, bulk-bought) goods. These were sold at only a modest profit, and the sale of alcohol was severely restricted. Largely by his own example, Owen began encouraging the cottagers to keep their streets clean. Initial hostility to the new manager amongst the workers gradually gave way to a long, undeclared truce of wary suspicion.

Then in 1807 America embargoed all exports to Britain. With no cotton available, the mill was forced to close down for four months. Owen desperately sought out alternative sources, whilst retaining the workers on full pay. His partners in Manchester were outraged, but the workers now undersood that Owen genuinely had their interests at heart. However, the greatest improvement was the opening of a school. This was financed by profits from the company store. No children under the age of eleven were allowed to work in the factory. Instead they attended school. Owen's wife acted as a teacher, along with other women from nearby villages; and an old soldier was brought in to teach the boys drill. A contemporary illustration of the school interior shows a light, airy, high-ceilinged room with tall windows. Three musicians are playing fiddles, and the girls are dancing in groups, while some visitors look on. Above them on the walls are large illustrations of animals: a giraffe, a lion, an elephant and the like. Owen paid particular attention to the school. He was convinced that early upbringing forged character, forming an outlook and good habits that could last a lifetime. In his view, 'Man is the creature of circumstances.'

Robert Owen, whose idealistic schemes for
self-supporting communities brought hope
for thousands

He even opened the first infant school in the land in 1816, which took in children from the age of one. Infants were encouraged to dance and sing, in preparation for their later education. At the next stage they would not only be taught how to read and write and do simple arithmetic, but also to think for themselves and try to understand how the world worked. (All this at a time when children of every class were treated as small adults; education was only considered worthwhile for upper-class children; and no link was seen between environment and character.)

Despite borrowing heavily, especially from his father-in-law, Owen still managed to make the factory turn in a profit. For the most part his workers backed him enthusiastically and were open to new efficiencies. But still his partners continued to plague him. The profits were not as high as expected, and some heavily disapproved of his philanthropic experiments. In line with the pessimistic theories of Malthus and Ricardo, they felt sure that such practices could only lead to disaster. Owen was determined to show otherwise, and decided it was time to write down his views. These would not only publicize what he was doing, but would also show how a business *could* be run. His intention was to discourage competing employers from undercutting prices by simply exploiting their workforce and employing child labour. There was another, more humane way – which actually worked. As distinct from Saint-Simon, the moral element in economics would be more than some futuristic idea.

In 1813 Owen published two essays, which a year later would form the body of his book *A New View of Society: or, Essays in the Principle of the Formation of the Human Character*. Set down in print, Owen's views can appear heavy-handed. He was convinced that 'any general character, from the best to the worst, from the most ignorant to the most enlightened, may be given to any community, even to the world at large, by the application of the proper means'. The evils of society only came about when individuals chose to create their own character. Such choice had to be eradicated where 'the masses' were concerned. Free will was seen as 'this hydra of human calamity, this immolator of every principle of rationality'. Human beings were motivated by 'the happiness of the self', but this could 'only be attained by conduct that must promote the happiness of the community'. This is belief in social engineering in its most extreme and authoritarian form. The twentieth-century experience of both fascism and communism has taught us to be more than wary of such ideas. Yet there is no doubt that Owen had his heart in the right place. His actions were directed at amelioration, rather than the attainment of a specific utopian dream. He wished to help society get a coat on its back, rather than into a straitjacket.

This is seen in his more practical recommendations. Owen was prompted by the conundrum that widespread poverty should coexist alongside the growing abundance of the Industrial Revol-

ution. New scientific discoveries had resulted in increased mechanization, which had brought about increased productivity, but
this had resulted in widespread unemployment and suffering. In
1811 Britain had witnessed a rash of Luddite riots, where groups
of workers had broken into factories, destroying the machinery
that had put them out of work. This increasing unrest had only
been halted when the prime minister, Lord Liverpool, had brought
in harsh, repressive measures. The result had been a mass trial at
York in 1813, after which many of the ringleaders had been hanged;
while hundreds of others had been sentenced to 'transportation
beyond the seas'. (This quaint term meant the prisoners were
shipped to the far limit of the known world: the brutal, isolated
penal colony at Botany Bay in Australia.)

Owen argued that, although mechanization increased production, it should not be allowed to bring about 'a most unfavourable disproportion between the demand for and supply of labour'.
Such a state of affairs was self-defeating. It only led to unemployment, which led to a decrease in consumption, which resulted in
a vicious spiral involving yet further unemployment. What was
needed was a 'profitable market', and this would only be achieved
when labour acquired 'its natural and intrinsic value, which would
increase as science advanced'. This required a labour theory of
value, with goods valued according to the amount of time and
labour which had been required to produce them.

By now Owen's partners had lost all patience with him. But
word had begun to spread about his new philanthropic experiment.
This enabled him to interest a number of backers, who helped
him to buy up New Lanark. The new stockholders in the company,
who included Ricardo's mentor, the philosopher Jeremy Bentham,
asked only for a 5 per cent return on their investment. Over the
next few years Owen's work at New Lanark began to attract
increasing attention. In the decade following 1815 over 20,000
people were to visit New Lanark, including Prince Maximillian of
Austria and the future Tzar Nicholas I of Russia.

In 1815 Owen began lobbying parliament to pass legislation for factory reform. He argued that factories were more than just instruments for the creation of wealth. Their conditions played a formative role in moulding the character of those who worked in them. In other words, they created human beings as well as wealth – and should be judged accordingly. But the time was hardly auspicious for reform. By the following year Britain was plunged into the economic crisis caused by the end of the Napoleonic Wars. The government no longer required a wide range of goods and supplies to maintain the struggle against France. Unemployment was rife, in many parts of the country people were all but starving, and groups of soldiers demobbed from the army were roaming the highways and byways with no means of support. As demand plummeted, agricultural wages were cut in half, and contemporary statistics suggest that almost half the population was reduced to a state of destitution. Feelings ran high, 'Bread and Blood' riots began breaking out, and an atmosphere of fear gripped the land. Had the French Revolution been defeated, only for it to break out at home? By enacting violently repressive measures, the government just manged to hold the lid on – until gradually the economy began to pick up again.

The government was now in the mood to listen to measures for reform. In 1819 it passed the Factory Act, which contained many of Owen's ideas – though these had all been heavily watered down. Child labour was limited, but only in certain industries. Despite this, two vital precedents had been established. For the first time the government had stepped in to regulate how businesses should be run. At the same time, the government had also accepted that it was responsible for those who were too powerless to protect themselves.

Owen now turned his mind to the continuing problem of unemployment and came up with some highly original ideas. He suggested that instead of paying out dole and benefits, the government should finance the setting up of 'villages of co-operation'.

These would each house just over 1,000 unemployed people on 1,000–1,500 acres. They would be largely self-supporting, providing goods and produce for their own subsistence – only buying in such necessities as they could not create for themselves. The funds to purchase these outside necessities would come from the sale of any surplus goods produced within the village. But this was more than just some airy economic fantasy. Basing his ideas on his experiences at New Lanark, Owen even drew up a blueprint for the layout and housing of each village. The inhabitants would be housed in four rows of buildings which enclosed a central square. Each family would have its own separate home within these buildings, but there would be a central communal kitchen and dining hall. Children would live at home until the age of three, when they would be brought up by the community. As soon as the first few villages of co-operation had been established, their superior qualities would soon become apparent to all. Further villages would be established throughout the land, and even beyond. Then 'unions of them, federatively united, should be formed in circles of tens, hundreds and thousands'.

It is easy to recognize in such ideas the seeds of practical socialism, as well as the inspiration for the co-operative movement. However, Owen's contemporaries saw things differently. His ideas were greeted with howls of opposition from across the political spectrum. Reformers saw them as simply a way of getting rid of the poor, dumping them in dustbin 'villages of paupers'; conservatives detected a dangerous precedent for working-class autonomy. These villages of co-operation may appear irredeemably idealistic and impractical to our eyes, but this would be a misjudgement. Owen's ideas survive almost intact on the modern Israeli kibbutz.

By 1824 Owen had become exasperated at the lack of government action, and decided it was time to put his ideas into practice. He sold up his interest in New Lanark and set sail for America. Here, he puchased 30,000 acres on the banks of the Wabash River in Indiana. In 1826, on 4 July, he founded the village of New

Harmony and issued his Declaration of Mental Independence. This declared independence from private property, from irrational religion and from the bonds of marriage.

For Owen, marriage was an 'unnatural crime [which] destroys the finest feelings and best powers of the species, by changing sincerity, kindness, affection, sympathy and pure love into deception, envy, jealousy, hatred, and revenge. It is a Satanic device . . .' Despite this, Owen's marriage appears to have been thoroughly orthodox, exhibiting all the traditional traits associated with this type of permanent union. According to Owen's biographer, Margaret Cole, by the time his wife was 'in her middle age [she had] only half a husband, and . . . Owen was not so much neglectful as apt occasionally to forget that she existed at all'. He simply had more important things to do: such as air his views on the 'slavish superstition' of marriage.

Within months 800 settlers had poured into New Harmony. Unfortunately, in the words of J. K. Galbraith, it 'attracted some of the most accomplished malingerers in the American republic'. No one wanted to do the dirty work (or in many cases, any work at all). The community was soon rife with 'idealistic disputes'. These ranged from how the place should be run to who should be running it – and every explosive topic in between: politics, sex, religion and so forth. Meanwhile one enterprising member appropriated a plot of land and set up his own hootch distillery, which quickly established a thriving trade. Soon neither the community nor many of its thousand members were able to support themselves – either financially or literally.

After two years Owen had had enough of New Harmony. He divided up his investment between his three sons and one of his partners, in the process giving away over 80 per cent of his savings, and sailed back across the Atlantic. Back in Britain followers of Owen were to set up three further communities: one at Queenswood in England, one at Ralahine in Ireland and one at Orbiston in Scotland – but all would fail within a few years. Yet paradoxically, Owen himself was not a failure; nor was he regarded as such, either by others or by himself. Regardless of his personal experience – whether of marriage or other disillusioning human behaviour – he remained immune from self-doubt. On his return from America he was welcomed as the champion of the workers' cause. By now his ideas, known as 'Owenism', had begun to spread through the working classes. Villages of co-operation may have failed, but the ideas behind them were very much alive. Groups began setting up self-governing workshops and co-operatives. In order to link up these workshops, and provide an outlet for their goods, a National Labour Exchange was established, with branches and stores throughout the country. This contained the seeds of an even more ambitious scheme: an attempt to socialize money, by linking it directly to labour. The National Labour Exchange issued its own 'Labour Notes', which had values of one, two or five hours' labour. These could be redeemed for goods or clothes at Exchange Stores in most large towns. So, instead of something having a market value (expressed in terms of money), it would now have a labour value. Here was a revolutionary scheme to cut out the middle of the 'work = money = goods' equation. Fluctuations in prices would thus seemingly be ameliorated, or even eliminated. Yet despite the good intentions behind this arrangement, it remained in essence barter. Not a small step into a hitherto unexplored world, but a giant step backwards for mankind.

Love of money may be considered the root of all evil, but money itself is not evil. It remains civilization's greatest invention – far greater even than the wheel or the harnessing of electricity. The first coins appeared in seventh-century BC Lydia, on the west coast of modern Turkey, later ruled by Croesus, whose fabled wealth was largely created by this new invention. Despite being frequently debased, the institution of coinage has shown a resilience unmatched by almost any other human institution, apart from the similarly debased institution of marriage. The most ingenious thinkers throughout history have been unable to agree precisely what money is or how to define it, what to do with it or where it goes – but nonetheless it's still here. It *works*. Attempts to eliminate the exchange of money – even in restricted communities – are usually doomed to failure. Why? For the simple reason that money can be virtually anything. Like water in the hand it trickles free, escaping all our attempts to grasp its shape. It evades definition: from the ambiguous profundity of the Roman emperor Vespasian's 'Money has no smell' to the precise poetry of James Buchan's 'frozen desire'. It is always something more.

In 1833 Owen began campaigning for a reduction to an eight-hour working day, causing widespread apoplexy amongst the class who did no work at all. The following year Owenites founded the Grand National Consolidated Trades Union, launching the working-class trade union movement, which soon began sweeping the country. The movement embraced a mish-mash of idealistic, humanitarian and often unrealistic reforms, as well as a few penetrating economic insights. As Smith had understood, in a capitalist market economy the workforce itself represented a capital value. Owen took this to its logical conclusion: if this was the case, workers should be looked after like any other capital asset.

Within a year of its foundation the Grand National Consolidated Trades Union had a membership of over 500,000 – a massive number representing nearly one in ten of the working population. But the movement had little cohesion. Expectations were high, often unrealistic and often contradictory. Local union branches were unable to constrain the actions of their members. Spontaneous strikes damaged the standing of the central governing body, which itself soon fell into disarray. Christian members were outraged by Owen's apparent atheism and his views on marriage. His call for an end to private property was seen by many as a call for revolution. Some were in favour, others against – but both had misunderstood Owen. In fact, he remained a gradualist throughout his life. He was all for radical social transformation, but only step by step. Others would call for revolution and write fiery pamphlets; Owen insisted upon seeing how his ideas worked in practice. The irony was that, where he succeeded in his business enterprises, his political enterprises had a distressing tendency to fail.

As the trade union movement swept the country it produced a vicious backlash – 1834 was also the year of the Tolpuddle Martyrs. Six meagrely paid farm labourers in the village of Tolpuddle in Dorset had the temerity to form their own trade union branch. For this they were hauled before the courts, and all summarily sentenced to seven years' 'transportation beyond the seas'. The treatment of the 'Martyrs' provoked demonstrations throughout the country, and the sentences were later remitted.

In the same year the Owenites suffered another catastrophe. Amidst growing disagreement Owen resigned from the Grand National Consolidated Trades Union in the summer of 1834, and a month later it collapsed. With it went the National Labour Exchange, the Exchange Stores and the attempt to 'socialize' money. This was the end of the first wave of the labour movement. The battle had been lost – but not the war. Young men who had joined the movement would become the leaders of the next wave, the Chartists.

Meanwhile Owen continued to tour the country, delivering speeches on education, rational thought and marriage reform. To the working class he remained an ambiguous figure: for them, but not of them. Preaching wholesale social transformation, yet opposing revolution and urging restraint. In many ways, he remained the somewhat despotic paternalist who had run New Lanark, the fervent anti-marriage preacher who accepted his own marriage with apparent indifference. In the end he became that most Victorian of entities, a national institution. In 1839 he would even be introduced by the prime minister, Lord Melbourne, to Queen Victoria. (Outraged at the prospect of this social encounter, the Bishop of Exeter demanded of Lord Melbourne an assurance 'that he would not expose his Sovereign and himself to the reproach of having abandoned the best, the most sacred, and the most holy interests of mankind'.)

In 1844, at the age of seventy-five, Owen returned to America for a prolonged visit, joining the younger members of his family at New Harmony. His sons had transformed the village into a going community. (New Harmony would last as an Owenite community until late in the twentieth century; to this day it is run along quasi-communal lines.) Owen's son John Dale Owen had gone on to found a similar community for freed slaves near Memphis, Tennessee. By the time Owen returned to America, his son John had been elected to represent Indiana in Congress, where he proposed a measure for settling the Oregon border dispute with the British in Canada. At the time the disputed Oregon Territory covered the entire northwest corner of the modern USA (including all of Washington and Oregon, as well as sizeable chunks of Idaho, Montana and Wyoming) and much of what is now western Canada. When Robert Owen returned to England he used his influence with the government to get them to accept his son's proposal. In this way, around one-tenth of mainland USA, and nearly half its northern frontier between the Lakes and the Pacific, came about as a result of the Owens.

When Owen had turned eighty he became converted to spiritualism. He continued to agitate for social reform and 'the dawn of reason ... when the mind of man shall be born again', but in the evenings he preferred having conversations with the ghosts of Benjamin Franklin, Shelley and the Duke of Kent. During his last years he returned to his home town of Newtown in the Welsh mountains, where he wrote his autobiography. He died in 1858 at the age of eighty-seven.

In Owen economics had found a humane answer to the laisser-faire pessimism of Malthus and Ricardo. His schemes may in the end have proved largely impractical, but they gave hope to many hundreds of thousands who might otherwise have been left in despair. The socialist tradition in Britain – which gave rise to the world's first comprehensive welfare state – stems directly from his ideas. Owen's experience of running New Lanark gave him a practical understanding of what economics meant. His idea that the workforce should be treated as a valuable capital asset was way ahead of its time. Morality and efficient capitalism can be the same thing. Only now are the implications of this concept being fully explored. The Indian economist Amartya Sen, who won the Nobel prize in 1998, was early in his career advised to avoid such 'ethical rubbish'. Ignoring this advice, he showed how educating an illiterate population effectively revalues a nation's assets, thus increasing its capital value. He also showed that such considerations are not limited to the Third World (which was what Britain, the most advanced industrial nation on earth, was like in Owen's time during the 1830s). With regard to the workforce, ethics and commercial worth are inextricably linked, especially where unemployment is concerned. Simply doling out welfare actually degrades the unemployed who receive it, and is thus wasteful of a nation's capital twice over. Besides money, the unemployed need to remain integrated in the community, so that they retain their sense of self-worth. This worth is part of the nation's capital. Ruined human lives are part of the national ruin. Such talk is not metaphorical. Sen has shown how this makes sense in hard economic terms. In the beginning the trade union movement sought a better life for its members. By the time of the late-twentieth-century monetarist era the unions themselves had become monetarist: the fight was largely for money. Sen's ideas are a direct development of the economic insights stemming from the era of the early trade unions. Like them, he seeks the development of human potential, which leads to well-being. The real direction of economic growth is the development of human ability. Any historian would find this obvious. Many economists still do not.

7

The Pleasure Principle

All this brings us to the question of human happiness. What is it? And what part should it play in economic philosophy? In the early eighteenth century Adam Smith's inspiring teacher Hutcheson had coined the phrase 'the greatest happiness of the greatest number'. Just over half a century later the English thinker Jeremy Bentham would make this the basis of his entire philosophy, which became known as Utilitarianism.

Jeremy Bentham was born in London in 1748 and quickly exhibited signs of being a child prodigy. It was said that he could write out the alphabet before he could speak properly. His father, a prosperous lawyer, decided to devote his time to tutoring his son, encouraging him with gifts. When the young Jeremy mastered Latin at the age of five, he was rewarded with a pink waistcoat. This practice gave him much pleasure, and was to prove a profound influence on his thought. By the age of sixteen Bentham had already graduated from Oxford and was studying law. Bentham Senior harboured dreams of his genius son rising to the top of the profession and becoming lord chancellor. But Bentham was averse to practising the law. He was more interested in the purpose of legislation. Instead of becoming a lawyer, he chose to bury himself in books. His widespread reading soon began to focus on philosophy and social reform. In 1776 he published *A Fragment on Government*, the first work of philosophical radicalism in the English language. This advocated such measures as birth control, the legalization of unions and votes for all (including women). For the most part these reforms would not be adopted in the following century, but the century *after* that.

Unlike the visionary dreams of Robert Owen, Bentham's ideas were backed by a fully reasoned moral and social philosophy. This was Utilitarianism, which Bentham based upon what he called the 'sacred truth'. In his own words, 'The greatest happiness of the greatest number is the foundation of morals and legislation.' His aim was to make the principles of the social sciences, including economics, as rigid as the laws of natural science. Utilitarianism and its pleasure principle would be the new gravity of morality, and he would be recognized as the Newton of social science. Bentham's basic argument was clear enough: 'Nature has placed mankind under the governance of two sovereign masters, *pain* and *pleasure*.' This was the 'principle of utility', which was a moral principle. What gives us pleasure is good, what gives us pain is evil. But for such a principle to be moral it must be viewed in social terms. What is right maximizes the pleasure of all, what is wrong causes an overall increase in pain and suffering. When confronted with difficult decisions we must weigh up the net pleasure against the net pain.

This brings us to the main difficulty of Utilitarianism. How is it possible to measure pleasure – either on the individual or the collective scale? Bentham tackled this problem in some detail, coming up with his 'felicific calculus' for the precise measurement of pleasure. In his analysis he listed seven different dimensions of pleasure – including its duration and the number of individuals affected by it. He also listed fourteen different types of simple pleasures – ranging through those resulting from power, wealth, skill, good name and (last but not least) malevolence. He also named a dozen 'simple pains' – ranging from disappointment to desire (a category which would seem to render most of us mashochists). But the plain fact is that pleasure, whether solitary or social, remains beyond precise quantification. And this remains so, even now that we can investigate it at the biological level. There is no fixed scale by which intensity of stimulus can be universally related to consequent enjoyment. An Indian fakir may enjoy a fiery curry, or a bed of nails, which a Dane finds intolerable.

But what has pleasure to do with economics? Very little indeed for many who suffer at its hands. As a result of the Utilitarians, the notion of 'utility' became central to economic thinking. Prior to Bentham a number of economic thinkers had used this word in general terms – often with reference to the 'desiredness' of goods or services. Bentham wished to give this central concept a precise and calculable meaning. He succeeded, in his own eyes, and 'utility' entered the economic canon, where it has remained to this day. Despite this, and despite the fact that the term remains in constant use (both in discourse and precise formulae), it still eludes ultimate precision. Modern definitions of utility refer to 'the welfare or satisfaction gained from the consumption of goods or services'. They point out that utility can be measured by prefer-ence, and can be said to increase with quantity. Yet in both cases this quantification is relative, with no absolute yardstick. Ultimately, the difficulty remains: human emotion is not suscep-tible to precise measurement.

Nonetheless, Bentham's analysis has given rise to an important economic tool, which has particularly frequent application with regard to government policy. This is 'cost-benefit analysis'. According to Bentham, the *benefit* which results from government spending should be measured against the *cost* of 'most vexatious and burthensome tax'. Cost-benefit analysis is now regarded as a systematic science, giving rise to a plethora of graphs (whose benefit is often far outweighed by the cost of trying to understand them). However, at the heart of this mathematical exactitude remains the same difficulty. 'Benefit' must be reduced to concrete goods or services, which are no precise measure of happiness.

Bentham was a bookish man, who preferred the solitude of his study to the company of others. He inherited enough to live comfortably, without having to worry about getting a job. As a result, while he wrote obsessively, he seldom got around to prepar-ing his works for publication. Despite his retiring personality,

Bentham was regarded as leader of the Philosophic Radicals, an active pressure group which lobbied parliament for reform. Utilitarianism provided a handy test for any new legislation. Would it bring about the greatest happiness for the greatest number? The Utilitarian principle rendered transparent a government's motives, and continues to do so. Is it acting for the general good, or simply for the good of its own vested interests? The Philosophic Radicals played a significant role in the passing of the Great Reform Act of 1832. This increased the electoral register by 50 per cent and abolished the rotten boroughs (parliamentary seats, such as Ricardo's, which could simply be bought). Even so, only one in thirty of the population now had the right to vote. The Philosophic Radicals, and their concern with social justice, were just the beginning of a long uphill struggle for political representation.

Bentham's concern with public matters meant that he was not entirely able to live the solitary life he preferred. He met Ricardo, whose astute judgement he admired. When Owen approached Bentham to persuade him to buy a share in the New Lanark mills, Bentham consulted Ricardo before investing. For some years Bentham's constant assistant was Ricardo's friend and mentor, James Mill, to whom he gave generous support. Bentham allowed James Mill and his family to live in the garden house of his London home at Queen's Square. (This dwelling had once been occupied by the poet Milton, who, according to persistent legend, was involved in the clandestine transfer of Oliver Cromwell's cadaver and its burial in the garden.) Mill and his family would accompany Bentham when he went to stay at Ford Abbey, the large, remote country house he rented in Devon. Bentham and Mill would work together in the ancient panelled library, debating philosophic points in the evening beneath the tapestried walls, and for exercise they would play shuttlecock in the great banqueting hall. Eventually James Mill's son John Stuart Mill became interested in Utilitarianism, and at the age of nineteen he edited a vast batch of Bentham's papers into a five-volume work, *On Evidence*.

Bentham's comparatively few published works were translated into all major languages. They covered a wide range of subjects, from philosophy to the police, from constitutions to prisons, with a constant emphasis on Utilitarian aims, practicality and reform. Indeed, his works had a far wider effect across Continental Europe and North and South America than they did at home in Britain. The French statesman Talleyrand visited him; his constitutional code was considered for the newly independent Greece; and Simon Bolivar, the liberator of South America from the Spanish, was fulsome in praise of Bentham during his mission to London – though he took the precaution of banning Bentham's works when he later became president of Colombia.

Bentham may have been a shy, retiring character, but he was not without vanity. Before he died in 1832 he issued strict instructions on how his body should be preserved, dressed in his own clothes and kept seated in a glass case. To this day this somewhat ghoulish sight remains on public display at University College London.

Shortly after editing Bentham's *On Evidence*, the nineteen-year-old John Stuart Mill had a nervous breakdown. This was not simply a result of piecing together thousands of scraps of paper covered with scrawled handwriting into a consecutive manuscript of over a million words, though such a task can't have been conducive to mental health. John Stuart Mill's breakdown was in fact almost entirely attributable to his father. James Mill, the mentor of Ricardo, the friend of Malthus, the intellectual partner of Bentham, who was renowned for his humane, compassionate approach in philosophy and economics, was in fact a classic Victorian monster in the privacy of his home. In emulation of Bentham's father, Mill Senior decided that he would educate his son to become a genius. From the very start he imposed total control on John Stuart's life in order to achieve this. At three the child was learning ancient Greek, by seven he had read most of Plato, and by thirteen he had in his own words finished

*Jeremy Bentham's mummified body in
its glass case at University College
London*

'a complete course in political economy'. This was no exaggeration:
he would now sit in on discussions between his father and Ricardo,
absorbing the very latest economic ideas. In 1821 the fifteen-year-
old Mill came across Bentham's *Treatise on Morals and Legislation*. By
the time he had finished reading through this three-volume work, 'I
had become a different being'. His admiration for Bentham and his
Utilitarian ideas knew no bounds. 'From now on I had what might
truly be called an object in life; to be a reformer of the world.' Three
years later he started into the mammoth task of putting together
Bentham's *On Evidence*.

However, all was not well with the young man: such forced precocity was unnatural, and would take its toll. As Mill later tellingly remarked in his *Autobiography*, 'I was never a boy.' His father forbade him any holidays, 'lest the habit of work should be broken and a taste for idleness acquired'. One can but imagine the domestic atmosphere presided over by a man described by his son: 'for passionate emotion of all sorts . . . he professed the greatest contempt. He regarded them as a form of madness.' But what did young Mill's mother make of all this? Either through fear, weakness or resignation, she did nothing. She had evidently seen it all before: her mother had run an asylum. Indicatively, John Stuart Mill does not mention his mother once during the entire 325 pages of his *Autobiography*. So succesful was James Mill's brainwashing of his son, that by the age of twenty even John Stuart's inner life was governed utterly by reason. He could allow himself no deviation from the tyranny of his indoctrinated mind. He even seems to have remained unaware of his exceptional talents, judging them to be 'below rather than above par, what I could do could assuredly be done by any boy or girl of average capacity'. Such a judgement could only have come from someone who had simply never met – let alone freely conversed with – someone of his own age. Even so, a growing self-awareness gradually began to dawn in the young man who had never been a boy. Working obsessively as ever amidst the gloom of a foggy London autumn day in 1826, he found himself pausing to ask himself, 'Suppose that all your objects in life were realized . . . would this be a great joy and happiness to you?' In his own words: 'An irrepressible self-consciousness distinctly answered, "No!" At this my heart sank within me: the whole foundation on which my life was constructed fell down.' The young man who believed utterly in the philosophy of happiness was incapable of achieving this for himself.

John Stuart Mill had a nervous breakdown. Again typically, he appears to have kept this to himself. Even more typically, neither his mother nor his father seems to have noticed. Yet amidst the inner turmoil a profound sea-change was taking place. Mill began reading the Romantic poetry of Wordsworth, and irrational thinkers such as Saint-Simon. Then one day he found himself 'accidentally' reading the *Mémoires* of the sentimental French poet Jean-François Marmontel. When he came to the passage where the poet describes the death of his father, Mill burst into tears. He claimed that he was moved by his own 'vivid conception' of the scene, though he makes no further reference to what was actually taking place in it. His suppressed wish that such a scene should take place in his own life would not have been so transparent in those pre-Freudian times. Mill concluded that he was now cured: 'The oppression of the thought that all feeling was dead within me, was gone.' John Stuart Mill would now devote himself to the introduction of humanity into political economy.

Bentham had conceived of Utilitarianism, and had proposed that a 'felicific calculus' should be applied to the social sciences. Mill would extend this inexact notion, applying it to the entire scope of human affairs. He would also make major contributions in fields ranging from logic to political theory, from philosophy to the emancipation of women. More relevantly, his profound insights into the workings of political economy would introduce Utilitarian considerations into a wide area where previously little notion of morality had prevailed. Adam Smith had been aware of injustices, and had pointed them out. Robert Owen had made his fortune in the free market, and then attempted to apply humanity within a different subsidiary system. Mill would indicate how the pleasure principle could be applied to mainstream capitalism itself.

Mill's major economic work was his *Principles of Political Economy*. This was to become the first generally accepted textbook on the subject, and as such would play a major role in the teaching of economics through the second half of the nineteenth century. In this work Mill drew widely, often without acknowledgement, on the economic work of another remarkable figure of this era: Charles Babbage. Today Babbage is best remembered as the neglected pioneer whose immensely intricate mechanical 'difference engine' displayed the hallmarks of the modern computer 150 years ahead of its time. Babbage was also Lucasian professor of mathematics at Cambridge, the illustrious post whose holders extend from Isaac Newton to Stephen Hawking. In 1832 Babbage published *On the Economy of Machinery and Manufactures*. In this work he displayed the technological understanding which he had acquired whilst visiting factories to gain the supreme engineering skill that enabled him to build the first purely *mechanical* (i.e. non-electrical) computer. Adam Smith had placed agriculture at the centre of the economy; Ricardo's great concern had been the grain trade. Babbage was the first to understand that the factory was now the central driving force of the economy. On his visits to factories the intent Cambridge mathematician soon grasped what business was about – rudiments which were not always apparent to the economic theorist perusing a description of a pin factory in his study, or discussing his latest ideas with a plutocratic intellectual financier while his teenage genius son sat across the drawing-room table. Babbage transformed Smith's notion of the division of labour, bringing it into line with the practice of the late Industrial Revolution. His analysis of the essential components of a factory, and how these worked together, was the origin of time-and-motion studies. His description of the economy included a definition, in economic terms, of the machine, as well as a conception of the vital role played by invention. He also understood that, where factory production was concerned, 'distant kingdoms have participated in its advantages'. The advance of technology and the spread of industry involved the entire world, which he believed would eventually benefit from this development. It is not difficult to see how Babbage's work would play an influential role in the thought of Marx as well as Mill – enabling them to come to their own widely differing conclusions.

In the *Principles of Political Economy* Mill in general follows the economic principles first set down by Smith, and later developed by Ricardo. This tradition came to be known as 'classical economics'. But Mill also made several important discoveries which would change this classical tradition for ever. The most profound of these was the most simple. Mill realized that the laws which govern economics are concerned with production, not with distribution. The productivity of labour, of the soil, of machinery – these can all be organized more or less efficiently according to certain objective laws. These laws are affected by certain limiting factors – such as nature (glut or famine), productivity (of labour, of machine) and so forth. The wealth is thus produced according to laws and objective factors which can enable us to maximize its quantity. But once it has been produced, there is no way we can 'maximize' its distribution.

Prior to this, Ricardo, Malthus and their followers had simply absolved themselves when confronted with the consequences of the economic laws which they discovered. The free market had its own 'natural' laws, and there was nothing that could be done about them. In a glut, prices dropped – and thus wages were bound to follow suit. As arable land became more scarce, the rent went up. It was as simple as that – no matter the suffering involved. Mill's uncoupling of production and distribution meant an end to such natural laws where distribution was concerned. Once wealth had been produced, once goods were ready for market, 'mankind, individually or collectively, can do with them as they please. They can place them at the disposal of whomsoever they please and on whatever terms.' Distribution had no 'natural' laws. 'The rules by which it is determined are what the opinions and feelings of the ruling portion of the community make them, and are very different in different ages and countries, and might be still more different if mankind so chose.' With distribution, morality became a factor. Ethics had at last found its way into capitalist economics. (Owen had sought to introduce morality by changing economics into a socialist system. Mill had introduced it into the very citadel of capitalism.)

Mill suggested that distribution was where Utilitarian principles

*John Stuart Mill, whose economic theory sought to
humanize classical economics*

could be applied. Yet he withdrew from the egalitarian consequences of pleasure being equitably distributed between the
owners and the workers. In economics Mill retained the classic
laisser-faire ideas of Ricardo. Mill believed that 'with the exception
of competition among labourers, all other competition is for the
benefit of labourers, by cheapening the articles they consume'.
This also dovetailed with his belief in unrestrained individual
liberty. Here he had but one proviso: 'The liberty of the individual
must be thus far limited; he must not make himself a nuisance to
other people.' Avaricious factory-owners who paid their workers

subsistence wages and housed them in slums were evidently not making a nuisance of themselves.

Mill's other great contribution was with regard to economic growth. This is all the more astonishing, when one considers his notion of what this was. Mill believed that economic growth was a mere passing phenomenon, a historical freak occasioned by the Industrial Revolution. Once this temporary hiccup was over, and a sufficient level of prosperity had been achieved, economics would return to the glorious stasis of the medieval era. Then instead of wasting their time scrambling to make ever more money, people could grow up and get on with the real things of life – such as thinking about liberty and social justice. Alas, it seems only the ancient Greeks were capable of this exalted state: when economics still meant discovering where the chickens had laid their eggs in the kitchen. As far as Mill was concerned, economic growth would grind to a halt as soon as the Industrial Revolution achieved its technological aims: the ultimate spinning jenny, a perfected steam engine, a completed network of railways and so forth. Such a view appears eccentric only with hindsight. The first railway covering less than a dozen miles between Stockton and Darlington in the north of England had opened in 1825. By 1841 there were over 1,300 miles of railway in Britain. It seemed such expansion was bound to come to an end.

Other indicators give an even more dramatic picture of growth through history in the region now covered by the Western European economies. It has been loosely estimated that the economy of this geograpical region approximately doubled during each of the following periods:

500–1500: the 1,000 years from ancient Greece to the Renaissance;
1500–1700: the 200 years from the Renaissance to the start of the Industrial Revolution;
1700–1800: the 100 years of early Industrial Revolution;
1800–1850 The fifty years of the late Industrial Revolution.

Mill's prognosis for economic growth could not have been more wrong. Yet ironically, his analysis of economic growth could not have been more right. Smith had seen growth as a benefit, bringing more freedom of trade and more wealth. Malthus (as well as a reluctant Ricardo) had foreseen only apocalyptic consequences such as overpopulation and mass starvation. Mill perceptively argued for both points of view. The interplay of such forces could bring about a number of possible results. Likewise, unforeseen factors such as natural disasters or labour unrest could also affect the result. Mill was the first economist to see that economic forecasting can never be certain. All the economist can do, *at best*, is indicate a number of possible outcomes. And given any amount of information, this will always be the case.

Even with the huge amount of pertinent data and accompanying computer power available today, economic forecasting can never be certain. This is because economics is subject to chaos theory – which accounts for how the flutter of a butterfly wing in the Brazilian rainforest can eventually result in a tornado in Kansas. In economics, a downward flutter in the value of the Thai baht on the Bangkok Stock Exchange one summer afternoon in 1997 eventually resulted in a financial hurricane which ruined the entire economy of the Far East, leaving a trail of devastation in markets around the world. Economics is subject to chaos theory because it is non-linear – that is, its variables are liable to increase at such a rate, in such a disproportionate manner, that any predictive formula quickly becomes incomputable.

It's all very well saying what economics *can't* do. But what about what it *can* do? In *On the Definition of Political Economy* Mill examines with philosophical rigour the entire question of economic method. Here, he explains how economics, in common with other behavioural or moral sciences, can never achieve the certainty of the natural sciences, such as physics or chemistry. This is because it is impossible to conduct meaningful controlled experiments in economics. Cause and effect cannot be precisely established, let alone measured. This is shown by the impossibility of finding two economic situations which are identical *except for a single factor*. Despite all this, Mill insisted that the physical sciences should remain the 'proper models' for economics. Here, we approach a crucial dilemma, which lies at the heart of economic theory. It isn't a science, yet it must try to behave like one. Faced with the essentially ungovernable complexities of the economic situation we can't simply give up, or give up taking economics seriously, just because we don't know precisely how to handle that situation. Economics may be less than physics, but it is more than astrology. The economic theorist's intellectual condition is (perhaps inadvertently) indicated in the following pronouncement by the great twentieth-century Austro-American economist Joseph Schumpeter: 'any objectionable piece of methodology is immaterial whenever it can be dropped without forcing us to drop any result of the analysis which is associated with it'. If only everything was like this: then no matter what we did, we need never be wrong. No wonder so many economists remain convinced – despite all evidence to the contrary – that they never are.

8

Workers of the World Unite

In the same year that Mill published his hugely influential *Principles of Political Economy*, Karl Marx published the first *Communist Manifesto*. That year, 1848, also saw revolutionary disturbances throughout Europe, from Sicily to Warsaw. In Paris the uprising led to the fall of the Orléans monarchy. In Vienna the reactionary and repressive Chancellor Metternich was forced to flee in disguise 'like a criminal'. After leading the unsuccessful uprising in Prague, the Russian anarchist leader Bakunin made his way to Dresden, where he befriended the young composer Wagner. A year later they would together take part in the abortive Dresden uprising. (Siegfried, the hero of Wagner's Ring cycle, is in part based upon the character of Bakunin.) Under cover of darkness Wagner and Bakunin managed to escape with their lives as the Saxon and Prussian troops crushed the rebellion. According to Clara Schumann, who remained in the city:

> They shot down every insurgent they could find, and our landlady told us later that her brother, who owns the *Goldener Hirsch* in Scheffelgasse, was made to stand and watch while the soldiers shot one after another twenty-six students they found in a room there. Then it is said they hurled men into the street by the dozen from the third and fourth floors. It is horrible to have to go through these things! This is how men have to fight for their little bit of freedom! When will the time come when all men have equal rights?

Far more than John Stuart Mill could have conceived possible,

188

Karl Marx got economics wrong, but also got it right. The communism he proposed to replace capitalism doesn't work; yet several of his most perceptive criticisms of capitalism remain unanswered. The questions of social justice which Marx raised – pressing and crucial at the time – still remain with us. The cheek-by-jowl existence of luxury and pitiless destitution found in Bombay and Rio de Janeiro would be all too recognizable to the Marx who inhabited Dickensian London. Even in the heartlands of affluence created by capitalism, its 'contradictions' are still evident in the urban ghettos of Chicago and Los Angeles, the economic wastelands of Northeast England and Naples. Capitalism has become *the* worldwide success story, but at a cost. In Marx's time, this cost was beginning to appear unbearable.

Like all the better class of revolutionaries, right through to Jean-Paul Sartre and Che Guevara, Karl Marx came from an impeccable bourgeois background. His father was a well-off lawyer; an uncle founded the Dutch industrial giant Philips. Karl was born on 5 May 1818 in the small provincial German city of Trier, amidst the picturesque vineyards of the Moselle valley close to the Luxembourg border. Two significant facts: Trier had recently come under the repressive regime of Prussia; and it was also in a rural region which had hardly been affected by the Industrial Revolution.

Both Karl's parents were Jewish, but had found it expedient to convert to Christianity for social and professional reasons. Besides practising his profession, Karl's father owned a couple of vineyards and led the life of a cultured liberal. He read Voltaire and joined a club which pressed for the autocratic Prussian state to adopt a constitution.

By the time Karl Marx was a teenager, he was dividing his time equally between the libraries and the taverns. As a result of his riotous activities in the latter he was challenged to a duel by a Prussian cadet, and was lucky to escape with only a traditional duelling scar. In 1835 he went to the nearby University of Bonn, where he continued reading voraciously and 'wild rampaging' (the words of his exasperated father). But after a year he transferred to the University of Berlin, ostensibly to study law – though by now he was only interested in philosophy. Here in the Prussian capital, well clear of the wine-loving Rhineland, student life was a more serious matter. The great Hegel had been professor of philosophy until his death just five years previously, and his followers, the Young Hegelians, were earnestly developing his philosophical ideas.

Marx diligently attended the lectures on Hegel's philosophy, but eventually fell ill 'from intense vexation at having to make an idol of a view I detested'. Yet Hegel would prove to be one of the two major influences on Marx. Hegel's philosophy saw the world and all history in terms of a vast, all-embracing, ever-evolving system. This evolution grew out of the struggle between contradictions, and worked in a dialectical fashion. Each notion implied the notion of its contradiction. For instance, the very notion of 'being' implied the notion of 'non-being' or nothingness. These two opposites (the thesis and its antithesis) then came together to form their synthesis, which was 'becoming'. In Hegel's all-embracing dialectical system, this synthesis then became a new thesis, which in its turn developed its own antithesis, and so on. This dynamic system moved through all ideas, all history and all phenomena – up to the highest level of absolute spirit reflecting upon itself, which was the totality of all that existed.

The second major philosophical influence on Marx was Ludwig Feuerbach, who had abandoned theology in order to study under Hegel. Feuerbach contended that God was merely the projection of man's inner nature at a certain stage of his development. Once this became clear, it was possible to transform Hegelianism. Matter was not dependent upon spirit, as Hegel had believed – it was the other way about.

Marx now began developing his own philosophy, which attempted to combine these two seminal ideas: Hegel's dialectical notion that everything evolved out of self-generating contradictions, and Feuerbach's insistence upon a materialism which saw ideas as a mere reflection of humanity's material conditions. But Marx's youthful passion would translate such ideas into heroic form. His doctoral thesis extolled Prometheus, the ancient Greek hero who stole fire from the gods and brought it down to humanity. For his punishment, Prometheus was chained to a rock in the Caucasus, where an eagle returned each day to peck out his ever-renewing liver. Marx would continue to identify with Prometheus throughout his life, and this ancient hero would provides an uncanny metaphor for the fate of Marx and his ideas. The Greek version of Prometheus means 'he who sees, or thinks, the future'.

When Marx left the University of Berlin, he had high hopes of taking up a post at some minor German university. Unfortunately his Young Hegelian mentor on the Berlin faculty was suddenly removed from his post as a result of pressure from the Prussian authorities. After searching somewhat haphazardly for a job, Marx eventually found himself a post as a journalist, working for the newly founded *Rheinische Zeitung* ('Rhineland Times'), a liberal newspaper based in Cologne. Marx turned out to be an excellent journalist, with a knack of coining memorable, ringing phrases. A year later he was promoted to editor. The idealistic, hard-drinking, hard-working boss was highly popular with his idealistic, hard-drinking, hard-working young staff, who nicknamed him 'Moor' on account of his swarthy, bearded features. The *Rheinische Zeitung* quickly became a thorn in the side of the Prussian authorities and its readership trebled, making it the highest-circulation paper in Prussia. Marx's social and political relationships now took an almost dialectical course, one which would remain characteristic of his whole life. Having attacked the authorities, he now proceeded to lambast the liberal opposition for its ineffectiveness. Next he launched into his left-wing staff, theoretical revolutionaries to a man, dismissing the whole idea of revolution as an impractical pipe-dream which simply hadn't been thought through properly. Despite such sentiments, the *Rheinische Zeitung* was closed down by the authorities in 1843.

Full of contradictions as ever, Marx soon decided to head for Paris, which had now become the centre of subversive activities, attracting would-be revolutionaries from all over Europe. But before he set off for France, the thirty-year-old Marx fulfilled the only long-term engagement he had so far managed to keep. For the last seven years he had been engaged to 'the most beautiful girl in Trier', who was an 'enchanted princess' (in the words of two of his rivals). Jenny von Westphalen was four years older than Marx, the daughter of Baron Ludwig von Westphalen, who came from a military family and held a senior post in the government administration. Jenny was intelligent and spirited, as well as being a local beauty. But the inane social life of Trier bored her to tears. Her marriage to Marx meant a romantic liberation.

Marx set off for Paris with his new bride, determined to join the revolutionary movement. One of his first actions was to become a communist, attending meetings of French working men's societies. Here again, the dialectic of Marx's inner contradictions soon came into play. His essentially academic mind found the ideas expressed at these meetings 'utterly crude and unintelligent'; yet his subversive temperament recognized that 'the brotherhood of man is no mere phenomenon with them, but a fact of life'. The individualistic intellectual and the rousing populist journalist synthesized to become a revolutionary. The full-blooded, but awkward, revolutionary realized that something had to be done, an intellectual programme had to be worked out. How was the revolution to come about? If politics was to change, then economics would have to change too. Marx began devouring the works of Adam Smith and Ricardo, digesting the concepts and dynamics of classical economics. In order to support himself, he secured an appointment as editor of the Paris-based *German-French Yearbook*.

It was through this magazine that he met a like-minded fellow-contributor called Friedrich Engels, whose father owned cotton mills in the Rhineland, and also one in England at Manchester. The 23-year-old Engels had been working for the family business in Manchester for the previous two years, but during the evenings devoted himself to pursuing his revolutionary ideals, meeting Chartists and followers of Robert Owen, as well as attending communist meetings. Engels' contradictions, unlike Marx's, were lived on the surface. A rebel at home, he still joined the family firm. Despite leaving school at seventeen, he went on to acquire a working knowledge of twenty-four languages. Though he functioned as a respectable businessman and member of the cotton exchange in Manchester, he also lived quite openly with his working-class girlfriend, an illiterate Irish redhead called Mary Burns. It was Mary who led him through the Irish slums off the Oxford Road, no-go areas for all but their inhabitants. During the course of these visits Engels encountered the scenes which appeared in his ground-breaking work *The Condition of the Working Class in England*:

> Masses of refuse, offal and sickening filth lie among standing pools in all directions; the atmosphere is poisoned by the effluvia from these, and laden and darkened by the smoke of a dozen tall factory chimneys. A horde of ragged women and children swarm about here, as filthy as the swine that thrive upon the garbage heaps and in the puddles ... The race that lives in these ruinous cottages, behind broken windows ... or in dark wet cellars, in measureless filth and stench ... must really have reached the lowest stage of humanity ... in each of these dens, containing at most two rooms, a garret and perhaps a cellar, on the average twenty human beings live.

Amazingly, this was in the year or so *before* the Irish Potato Famine, when a million would die and many more would be forced to emigrate, spilling into such 'Little Irelands' all over Britain and North America. Yet when Engels was walking with a fellow businessman, and pointed out how these slums were a disgrace to Manchester, his colleague merely listened politely and then remarked to him on parting: 'And yet there is a great deal of money made here. Good day to you, Mr Engels!'

When Engels began sending articles to the *German-French Yearbook*, Marx was immediately struck by the force of his subversive ideas. Engels called to see Marx in Paris on his roundabout way home for a holiday. The communist *bon viveur* and the grubby, cheroot-smoking journalist soon found they had a lot more in common than their large beards. During Engels' ten-day visit, the two of them struck up an immediate and profound rapport that would last a lifetime. This would be the only friend with whom Marx would never quarrel. Engels, for his part, worshipped Marx – the word is hardly too strong. He would devote much of his time and money in support of his hero-friend, to say nothing of the emotional and physical energy involved in this exacting task. Although Marx was married, and now had a baby daughter, he was still living the precarious attic life of a poor student. This too would remain a permanent feature of Marx's life. As we shall see, this was due to something more than mere financial necessity. The lack of respectability or social responsibility involved appears to have fulfilled some unresolved psychological need. Marx would remain poor for the rest of his life, yet it would never be working-class poverty, with the accompanying extreme squalor and despair, such as Engels had witnessed in Manchester. Marx's poverty was always more that of the perennial student fallen on hard times – often extremely hard times, but recognizably that of the improvident 'gent' nonetheless.

In 1845 Marx's magazine published a poem by Heinrich Heine satirizing another respectable German who would choose to live openly with his redheaded Irish mistress. Only this time the German was King Ludwig of Bavaria. The *German-French Yearbook* was closed down, and Marx expelled from Paris. He left for Brussels, where the impecunious Marx family was increased to four, when Jenny gave birth to a son.

Engels followed Marx to Brussels, and the two of them joined the Communist League. On account of their previous journalistic work, Marx and Engels were given the task of writing the first *Manifesto of the Communist Party*. From the very beginning this was something of a misnomer. There was no communist party – the Communist League was merely one of several groups who referred to themselves as communists. Likewise, the idea of producing a manifesto was not to proclaim communist policy, but to establish precisely what that policy was! Members of the Communist League embraced a wide range of idealistic and utopian ideas – ranging from revolutionary socialism to anarchism. Marx and Engels were expected to hammer these various woolly and disparate ideas into some definitive shape. This they would succeed in doing beyond the wildest dreams of their sponsors. The *Communist Manifesto* (as it is now more popularly known) would eventually become one of the greatest worldwide best-sellers in the history of printing, along with the Bible and Shakespeare. There is no doubting that this forty-page document is *the* masterpiece of its kind.

Its opening is suitably dramatic: 'A spectre is haunting Europe – the spectre of communism.' In an early draft, Engels defined communism as 'the doctrine of the conditions for the emancipation of the proletariat . . . that class of society which procures its means of livelihood entirely and solely from the sale of its labour'. This would be achieved 'by the elimination of private property and its replacement by community of property'. According to Marx, whose ringing phrases dominate the document, 'The history of all hitherto existing society is the history of class struggles.' This has passed from the slave era, through the feudal era to modern bourgeois society, where the capitalists are able to dominate the proletariat because they own the means of production, such as the machinery and the factories. Surprisingly, Marx is the first to admit the unparalleled achievements of the bourgeois age. 'It has been the first to show what man's activity can bring about. It has accomplished wonders far surpassing the Egyptian pyramids, Roman aqueducts and Gothic cathedrals; it has conducted expeditions that put in the shade all former Exoduses of nations and crusades.' *But* – and here comes the full weight of Marx's analysis –

> It has pitilessly torn asunder the motley feudal ties that bound man to his 'natural superiors', and has left remaining no other nexus between man and man than naked self-interest, than callous 'cash payment'. It has drowned the most heavenly ecstasies of religious fervour, of chivalrous enthusiasm, of philistine sentimentalism, in the icy water of egotistical calculation. It has resolved personal worth into exchange value, and in place of the numberless incontestable chartered freedoms has set up that single, unscrupulous freedom – Free Trade.

The human richness of medieval life (such as still lingered in pre-industrial spots like Trier) had given way to the industrial urban nightmare (such as could be witnessed from student attic windows in Berlin and Paris). Humanity had been dehumanized. Individual freedoms had been harnessed to free trade – the very factor which, according to Adam Smith, allowed the invisible hand of the market to do its work, providing benefit for all. Here, for the first time, was a trenchant analysis which was diametrically opposed to classical economics. The victory of the proletariat would bring about the first classless society. This great future bore a resemblance to the utopias envisioned by Saint-Simon and Robert Owen. However, the 'social utopias' established by such socialist dreamers were in fact no more than 'reactionary sects'. In opting out of capitalist society, they detracted from the wider struggle to overthrow it. And reforms which attempted to remedy the defects of capitalism were similarly misguided. Capitalism could not be rescued from its inevitable collapse.

Despite this, Marx does offer a list of reforms in the *Manifesto*. Those such as progressive income tax, the abolition of child factory labour and free education for all children we now accept as the norm. Others such as the abolition of private property, the establishment of a state monopoly in such fields as banking, communications, transport and all means of production have been tried – and have failed, for the most part catastrophically. The *Manifesto* ends with its celebrated call to arms:

> Communists disdain to conceal their views and aims. They openly declare that their ends can be attained only by the forcible overthrow of all existing social conditions. Let the ruling classes tremble at a Communist revolution. The proletariat have nothing to lose but their chains. They have a world to win.
> WORKERS OF THE WORLD UNITE!

Marx hurriedly penned the final version of the *Manifesto* in January 1848. Although his words would have no leading effect during the immediate 'year of revolutions', there was doubt that he had caught the mood of the times. In the very same month a local revolution broke out in Sicily; the following month saw one in Paris, then it spread to Germany, the rest of Italy . . . Many (and not just the revolutionaries) were convinced that the era of capitalism was coming to an end.

Having despatched their inspirational revolutionary manifesto to the grateful Communist League, Marx and Engels now astonished their benefactors by turning their back on the cause. They both left Belgium and returned to the Rhineland. Here, Marx accepted the post of editor of a resurrected *Neue Rheinische Zeitung*. The paper's relaunch had been financed by a group of local bourgeoisie. Marx's editorials were soon denying the revolution, and instead advocating a collaboration between the working class and the democratic bourgeoisie. But Marx's tactical change of policy was to prove short-lived. In September 1848 King Friedrich Wilhelm IV dissolved the Prussian Assembly in Berlin. This was too much for Marx, who immediately advocated armed resistance to this suspension of democratic rights. He was arrested, but put on a bravura performance at his trial. He told the jurors that he hadn't been advocating revolution, merely a defence of the realm. The

King himself had been guilty of revolution. Such was the popular feeling of the moment that Marx was unanimously acquitted, and even thanked by the jury, amidst a cheering courtroom.

Meanwhile the fearful bourgeois backers of the *Neue Rheinische Zeitung* had withdrawn their support, but Marx managed to put out a final issue. This was printed in bright red ink, with Marx announcing in his editorial that his 'last word everywhere and always will be: *emancipation of the working class!*' The paper caused the expected uproar, and Marx was deported.

In August 1849 Marx arrived all but penniless in London, accompanied by his family, which had now increased to three small children, with Jenny pregnant once more. In a show of solidarity, he and Engels rejoined the Communist League, whose international headquarters were in London.

At the turn of the nineteenth century, London had become the first city in the world to top 1 million inhabitants, and since then it had mushroomed as alarmingly as any current shanty-besieged Third World metropolis. During the decade preceding Marx's arrival, the population of London had grown by 300,000 to 2½ million. This was the capital of the British Empire, the largest empire the world had ever known, soon to be described with geographic accuracy and overweening ambition as the 'empire on which the sun never sets'. London was the richest city in the world, and at the same time contained more misery and poverty than any other in the world. Gaslight had been introduced to the streets as early as 1812, and Brunel had completed the first tunnel under the Thames just over thirty years later. But the streets which benefited from these innovations contained the underworld brought to life by Dickens, where Fagins ran gangs of child sneak-thieves, debtors ended up in the notorious Newgate Gaol, and convicts lay in chains inside the rotting hulks on the Thames estuary awaiting transportation beyond the seas. Perhaps the most telling image is provided by the Crystal Palace, which was under construction when Marx arrived in London. This building was to house the Great Exhibition of 1851, in which 'all civilized nations' had been invited to take part. When completed, the Crystal Palace was over four times the size of St Paul's Cathedral, large enough to contain any modern skyscraper yet built lying on its side, and apart from its framework it consisted entirely of glass – truly one of the wonders of the Victorian world. Yet the main stock-in-trade for the hawkers outside on a hot summer's day was perfumed handkerchiefs, used to mask the stench from the raw sewage clogging the nearby Serpentine lake in Hyde Park.

For almost a year after Marx's arrival in London he and his family were to live a hand-to-mouth existence. They moved from one cheap lodging to another in the shabbier backstreets around Leicester Square favoured by the many Continental political exiles. As the sky was being lit up by the firework celebrations of Guy Fawkes Night on 5 November 1849, Jenny Marx gave birth to her fourth child, a son. Marx named him Guido (nickname 'Fawksey') after the conspirator who had tried to blow up the Houses of Parliament in 1605. A few months later the Marx family was cast out on to the street with their few sticks of furniture, and only rescued by the charity of a fellow exile. More lasting charity was now provided by Engels, who had given up his attempt to become a journalist. He had gone back to work for his father's factory in Manchester, at least partly so that he could support Marx. At the beginning of 1851 Marx and his family found more settled lodgings in two rooms on the top floor of 28 Dean Street in Soho (a building which now houses one of London's most fashionable restaurants). This was the beginning of Marx's decade-long period of oblivion – a time of spiritual and political isolation, supported by hand-outs from Engels, who was exiled 170 miles away in Manchester.

The Communist League was Marx's only consolation. His charismatic and genuinely endearing personality, along with his daunting and far-ranging intellect, made him a natural leader. But his supreme political skills were best adapted to small groups, such as the newspaper office and the committee room. He had to dominate: he disliked appearing at public meetings, or encountering intellectual peers who might cross swords with him. The Communist League soon fell apart amidst a welter of bickering and recriminations.

Marx's house in Dean Street was kept under permanent surveillance by Prussian police spies, one of whom even managed to inveigle his way in. His report has left us with the most intimate picture we have of Marx's life during this period:

> As soon as you enter his room, your eyes are so dimmed by coal smoke and tobacco fumes that it is as if you are blundering into a cave ... Everything is so dirty, and the place so full of dust, that even sitting down is a hazardous undertaking. The chair on which one sits has only three legs, the only whole chairs being used by the children to play and prepare food ... Besides being a poor host, Marx is also a completely disorganized and cynical person. He leads the existence of a geniune bohemian intellectual. He very seldom washes himself, combs his hair or changes his clothes. He also enjoys getting drunk. Sometimes he is idle for days on end, but he will work day and night with tireless endurance when he has a lot of work to do. He follows no routine when it comes to getting up or going to sleep. Frequently he stays up all night; then he lies down fully clothed on the sofa at noon, and sleeps until evening, oblivious to whoever passes in and out of the room.

In all fairness, this chaotic regime must in part have been imposed by the fact that Marx was living in two small rooms shared with a wife, three small children, their German maid Lenchen, and presumably the odd visiting Prussian spy taking outraged notes about the scruffy bearded figure snoring contentedly on the sofa in mid-afternoon.

Despite all this, Marx now set himself a tireless regime of research in the reading room at the British Museum. The 1848 revolution had failed, and a period of severe repression had set in over Europe, causing many radicals to despair. But Marx was equipped with exceptional psychological endurance. He would bide his time. Meanwhile he decided to work out his revolutionary ideas of political economy on paper, a task that would take him through the first long years of his isolation. By 1859 he had at last completed his first full-scale economic work, *Contribution to the Critique of Political Economy*. The starting point of Marx's thinking is the need for a fundamental change of attitude – towards both traditional philosophy and classical political economy. As he had said many years earlier, 'Philosophers have previously only interpreted the world; the real task is to change it.' Substitute economists for philosophers and we have the basis of his economic approach. Things had gone wrong: it was time human beings took charge of their own destiny. Despite such pronouncements, Marx was no empiricist. True knowledge was not based upon experience, but achieved through the critique of ideas. This philosophical attitude was inherited from Hegel. Indeed, most of Marx's basic concepts were inherited from previous thinkers. As we shall see, in economics he accepted the fundamental notions derived by Smith, Ricardo and Mill. It was his critique of these concepts which was both radical and original.

Marx's philosophy, both political and economic, is based upon the following analysis. Social life is founded on economic life, upon how things are produced within a society. Social relations are based upon economic relations. Above these rises a corresponding superstructure of laws and social consciousness which reflects the economic structure. In this way the ideological and intellectual life of a society is entirely determined by the way things are produced within it. In Marx's words, which had already begun to generate their own lumpen jargon, 'The mode of production in material life determines the general character of the social, political and intellectual processes of life. It is not the consciousness of men which determines their existence; it is on the contrary their social existence which determines their consciousness.'

The year 1859 also saw the publication of Darwin's *Origin of Species*. Ideas of evolution were very much in the air. Marx sketches a philosophical evolution of consciousness – which he sees as developing in a dialectical fashion, rather than through the survival of the fittest. Originally we lived in harmony with nature (thesis). It was only by opposing nature that we realized ourselves as human beings (antithesis). Out of this struggle was born our consciousness (synthesis). Similarly, the further evolution of human consciousness has remained inseparable from struggle. However, this evolution had now reached a state where it was fatally blighted. Adam Smith had harboured misgivings about the division of labour (as exemplified by the pin factory), but he had seen it as a necessary advance in economic efficiency and productivity. Marx saw it as destructive of the consciousness of all involved. Reduced to the endless repetition of a single, mindless task, the worker had no meaningful relationship with the final product he was helping to create. He (or she) lost all pride in his or her work and became 'alienated'. Those who worked in such a fashion were literally dehumanized. They became mere drudges – alienated from their work colleagues, from the community through which these alien products were dispersed and sold and even from their family, who bought the products. There was no single tangible product which the worker himself could claim to have produced, nothing in which he could have pride and through which he could identify himself in society. Market production on a large scale simply intensified this process with further specialization, piece work, and the contracting-out of different tasks in the overall manufacturing processes. In this way the worker became estranged from his 'true' social being.

The notion of private property, so essential to market production, only added to this effect.

> Private property has made us so stupid and partial that an object is only *ours* when we have it, when it exists for us as capital, or when it is directly eaten, drunk, worn, inhabited, etc., in short

utilized in some way . . . All the physical and intellectual senses
have been replaced by . . . the sense of *having*.

Instead of satisfaction on an individual and communal level, all
the worker received was money. Literally and metaphorically,
hard cash. In Marx's view, money had 'deprived the whole world,
both the human world and nature, of their own proper value.
Money is the alienated essence of man's work and existence; this
essence dominates him and he worships it.' When the production
and marketing of goods is motivated entirely by profit, social
justice and even basic human needs are disregarded. Such an
economic world, which finds its *raison d'être* solely in profit, results
in grotesquely distorted social relationships. This affects all human
activity in such a society. The political, intellectual, artistic and even
spirtual life all echo this method of production, which is justified by
financial gain rather than any other form of social benefit. Seen in
this light, history becomes completely transformed. Morality, the
law, even religion, do not evolve according to a history of their own.
Such consciousness, both individual and social, is dictated by economics, by what Marx called historical materialism. Material existence dictates our consciousness, not vice versa.

The history of the twentieth century would show how Marx's
answer to these problems went catastrophically wrong. Private
property, money, the profit motive and alienation would seem to
be fundamental to the present stage of our evolution. We make use
of them – as they make use of us. Alienation becomes heightened
individuality. On the other hand, Marx's analysis reaches far
beyond the early Victorian society to which it was applied. His
description of money worship, our attitude to private property,
consumerism and the pursuit of profit for its own sake are all
too relevant to the age of downsizing, provoked currency crises,
rocketing (and plummeting) high-tech share prices which represent no commercial reality and companies whose assets consist
of everything except the people who work in them. The apotheosis

– both Freudian and literal – of this lust for lucre is perhaps best illustrated by the fact that man's deepest penetration into the earth so far, some 3,777 metres below the surface at Western Deep Levels in South Africa, has been in the pursuit of gold.

Even in the hirsute Victorian era, Marx stood out from the crowd – a fact only accentuated by his thick German accent, which he did nothing to improve. The diet of bread and potatoes, the cheap cheroots constantly staining and fumigating his beard and lungs, the sedentary life and heavy drinking – these soon began to take their toll. Marx started suffering from painful boils, which would continue to rack his flabby frame until the end of his days. But others in the family were not possessed of such stamina, and two more of his children died in infancy. On many occasions the family were literally penniless.

As if all this wasn't bad enough, Marx also had an affair with the maid, Lenchen Demuth, and made her pregnant. Engels, a frequent visitor, selflessly took the blame upon himself. When Lenchen was delivered of a small, dark, hirsute child Jenny had her suspicions, but kept these to herself for the sake of the family. Years later, on his deathbed, Engels would reveal the truth to Marx's daughter Eleanor (known as 'Tussy'). Young Freddy Demuth would grow up to become a true member of the proletariat, working in an engineering factory in Hackney, in London's working-class East End. In old age he would live to see the ideas of both his 'fathers' come to fruition in the Russian Revolution and the establishment of the Soviet Union. Freddy's siblings were less fortunate. As if the succession of child-deaths was not enough, the Marx children who reached maturity seem to have been equally cursed. His oldest daughter Laura would commit joint suicide with her anarchist husband, while living in poverty in Paris. His favourite Tussy chose the same path after being rejected by her philandering lover, who even gave her the prussic acid which she drank, causing her excruciating death.

Yet times were not unrelieved misery *chez* Marx. On sunny Sun-

days the family would travel up to Hampstead Heath for jolly pic-
nics, with leap-frog and other party games afterwards. There is even
a farcical description of a drinking spree which Marx embarked
upon with some German friends. This ended up with a 'student
prank' – smashing some gas lamps with stones, followed by a dash
through the night streets to elude the chasing 'bobbies'. In temperament Marx would remain very much the perennial student.

During the 1850s Marx managed to secure a job as London
correspondent of the *New York Daily Tribune*, at the time the
world's biggest-circulation newspaper. Marx was contracted to
write twice-weekly commentaries on British and Empire news,
though these would frequently have to be dashed off by Engels
in order to meet the deadline. Despite Marx's regular income, his
correspondence with Engels regularly included desperate pleas
for further cash, citing the imminent arrival of the bailiffs, no food
in the house, and such. Marx was completely open in discussing
his political ideas with Engels, and this intimacy also extended to
his personal life – even going beyond financial emergencies to
include such details as the eruption of a boil on his penis, or how
he'd decided take things easy and remain indoors because he'd
pawned his only pair of trousers so he could buy some cigars. Karl
Marx or Groucho Marx? It's sometimes difficult to tell. To quote
some relevant economic statistics, in the manner which Marx himself so favoured in his works: he received £150 a year from Engels,
as well as £2 for each of his bi-weekly articles for the *New York Daily
Tribune*. Even at his very lowest ebb his annual income never
dropped below £200, while the income for a clerk during this period
was £75. Marx's outgoings were modest – his annual rent for Dean
Street was just £22, and Lenchen would have received £20 a year
(had she ever been paid, that is). Somehow the remaining £178
simply evaporated into the tobacco-clouded air, while Marx sat sun-
ning himself at the window in his underpants.

Despite their poverty the Marxes always had a maid – the long-
suffering Lenchen had orignally been a Rhineland peasant girl sent

over to look after them by Jenny's aristocratic parents. Regardless of their bohemian existence, Karl and Jenny appear to have retained certain pretensions. Marx remained steadfastly unwilling to stoop to actual labour – prefering to write at great length about the conditions of such activity. It was this which prompted his exasperated mother to comment, 'What a shame little Karl doesn't make some capital, instead of just writing about it.' Meanwhile Jenny Marx still insisted upon using her inherited title 'Baroness von Westphalen' – a fact which would invariably be overlooked in the future Soviet and Chinese hagiographies of Marx.

It was from Jenny's family that the Marxes eventually received the small legacy which enabled them to escape from grim Dean Street and move to the more genteel suburban poverty of Grafton Terrace in north London. Despite his unwillingness to provide materially for his family, Marx remained a much-loved paterfamilias – referred to by one and all by his nickname 'Moor'. Visitors were liable to find him on all fours giving 'elephant rides', as his children clung to his back, his hair, his beard, squealing with delight. It was during these years that Marx allowed his hair and beard to grow longer and longer, assuming a consciously 'promethean' appearance. In his own eyes, he was now 'writing the future'. Marx also seems to have found a (brief) solution to the problem of providing for his family. When Jenny received another small legacy, he wrote in a letter to a friend,

> I have, which will surprise you not a little, been speculating . . . especially in English stocks, which are . . . forced up to a quite unreasonable level and then, for the most part, collapse. In this way, I have made over £400 and . . . I shall begin all over again. It's a type of operation that makes small demands on one's time, and it's worth while running some risk in order to relieve the enemy of his money.

Since this is the only mention Marx makes of his new hobby, we can only assume that next time round it was not 'the enemy' who

*Karl Marx and his wife, Baroness Jenny
von Westphalen*

was relieved of his money. At any rate, Marx now returned with renewed vigour to his radical dissection of capitalism.

The long hours of toil in the reading room of the British Museum would eventually result in *Das Kapital* ('Capital'), the first volume of which was published in 1867. This was Marx's masterpiece – the work of an economic analyst only equalled by Adam Smith and John Maynard Keynes. *Das Kapital* investigates the mechanism of economics against the background of mid-nineteenth-century Britain. Here was the most advanced industrial economy in the world, and it appeared to indicate the future. At the time, this seemed a valid deduction. Both in capacity and efficiency British industry far outstripped all its competitors. The full extent of British supremacy is given by a summary of one of those statistical charts so beloved by Marx. This one deals with the cotton industry:

Average number of spindles per factory:
England: 12,600
France: 1,500
Prussia: 1,500

Average number of spindles per person:
Great Britain: 74
Prussia: 37
France: 14

Despite this vast supremacy, the British workers' conditions were appalling. A Poor Law doctor in Bradford made a list (included in full in *Das Kapital*) which showed that his patients were living *on average* a dozen to a room, with some over double this. A street with over 200 houses was liable to have less than forty primitive outside lavatories. Those who lived under these conditions worked long and hard. A skilled factory hand in Northern Ireland was required to work from 6 a.m. until 11 p.m. Monday to Friday, stopping at 6 p.m. on Saturdays. 'For this work I get 10s 6d [52p] a week,' the worker explained to the visiting factory inspector. All the statistics which Marx collected were from the official government reports in the British Museum: the capitalist system freely provided the evidence against itself.

This was the system whose workings had been outlined by Smith and Ricardo. Marx came to the conclusion that it couldn't go on like this. As with the statistical evidence in the British Museum, Marx used the words of Smith and Ricardo against themselves, employing their concepts and their logic to demonstrate the failings of the system they described. Smith had concluded that under free-market capitalism the interest of the capitalist coincided with that of society. The invisible hand ensured that ultimately the economy worked for the benefit of all. This was patently not the case. Marx also pointed out that the classical economics of Smith and Ricardo 'proceeds from the fact of private property. It does not explain it.' Private property was not a permanent feature, as any glance at history would show.

In the beginning there had been tribal property; next ancient communal or state property; then feudal or estate property (conferring social 'status' on its owners); thence had come the bourgeois notion of private property. But what underlay all this social development? As we have seen, Marx viewed history as a succession of class struggles. In ancient society the slave class struggled against the freemen; later the Roman plebeians struggled against the patricians; then the serf against his lord; the medieval journeyman against the guild master. 'Oppressor and oppressed stood in constant opposition to one another . . . an uninterrupted fight, now hidden, now open, a fight that each time ended either in a revolutionary reconstitution of society at large or in the common ruin of the contending classes.' Historical progress marched in a dialectical fashion. Each phase developed its own contradictions, which eventually resulted in the progressive synthesis of a new social system. Capitalism was simply another phase in this inevitable historic progress.

As capitalism developed, it too generated its own inherent contradictions. A free market led to a growth in competition. In order to increase efficiency and profits for his business the bourgeois capitalist invested in machinery. Small businesses which couldn't afford such capital investment were driven to the wall. This intensifying competition led to larger and larger enterprises dominating the market, until eventually a monopoly was established. Hence competition led to its contradiction in the form of a monopoly. At the same time, the introduction of machinery meant increasing unemployment. But this served to decrease the market – the unemployed had no wages to spend on goods. Yet at the same time this increased efficiency meant more and more goods were produced. More goods, decreasing market, decreasing profits. Thus, further contradictions within the system emerged.

If, on the other hand, there was a boom which resulted in full employment, the workers' wages were bound to rise, according to the law of supply and demand. There would be no pool of unemployed who could be drawn upon to work for lower wages. Higher wages would eat into profits. Either way, the capitalist's profits would inevitably dwindle.

These internal pressures arose within capitalism as a result of its own development. The result was a series of recurrent and ever-deepening crises. These would eventually lead to the final crisis which would bring about the collapse of the capitalist system.

According to Marx, capitalism was basically unjust. It relied upon the exploitation of the workers. This arose because the capitalists owned the means of production: the machinery, the tools and so forth. A cotton bale arrived at the factory door, and left as garments which could be sold for a higher price. In this way the worker in the factory added value to the goods. But he was not paid the full value that he had added. In fact, he was paid a subsistence wage, or little more, and the factory owner pocketed the surplus value as profit. This, according to Marx, was exploitation.

Marx was a firm believer in the labour theory of value. A product had a *real* value which could be calculated according to the amount of labour which had gone into its production. When machinery entered the equation, this was valued according to the amount of labour which had gone into *its* production. As we have already seen, the labour theory of value is at odds with the supply and demand values dictated by the market. Similarly, Marx's analysis of the manufacturing process severely misjudges the role of the capitalist. It is he who risks his capital when he sets up the enterprise in the first place, and for this he requires a reward to make his investment worthwhile. This is the driving force of capitalism. Enterprise – imagination, risk. As Mandeville showed, its motives – like any economic motive – are not Christian. They are the acceptable face of avarice. No one embarks upon an enterprise if there is no possibility of gain, and the only prospect is the risk of loss. Such is human nature.

There are also other flaws in Marx's analysis. The surplus value which the capitalist 'appropriates from the worker' is not all profit. Admittedly, Marx does take into account the cost of depreciating machinery. But he pays only passing heed to the very element which he had indicated was endemic to capitalism – notably, expansion. If the capitalist was to survive in a competitive market he had to expand. This too had to be financed out of the surplus value.

The domineering and exploitative behaviour of the capitalist class – the demonized 'bourgeoisie' – in Victorian Britain was often grotesque. Likewise, their attitude to the hideous poverty they inflicted on the proletariat. ('Good day to you, Mr Engels!') In Marx's view, the balance of social and economic justice would only be redressed when the means of production were taken over by the state. Such forms of bourgeois private property should be nationalized. This is precisely what happened in the Soviet Union, and throughout the communist world. Free enterprise was stifled in favour of state planning: the five-year plan, the great leap forward, and the like. Under favourable conditions this may appear all very rational and just. But human evolution – either social or individual – has at best only aspired to reason and justice, rather than embodied these qualities. A controlled economy may attempt the occasional great leap forward, but it is unlikely to create a Silicon Valley. Such leaps spring from the intoxication of individual imagination rather than sober committees.

Even so, elements of Marx's critique of capitalism still have relevance. Many of these we choose not to notice. We prefer our own 'working-class heroes' to the Stakhanovite versions depicted six times life-size bearing red banners in socialist realist murals.

Yet the working-class heroes of contemporary capitalism only appear to have bucked the system. Our rock stars, football millionaires and whizz kid City dealers still don't own the means of production. Here today, gone tomorrow – while the owners of the means of production continue to pocket any surplus value. (The film stars of Hollywood understood this as long ago as 1919, when Chaplin, Fairbanks and others created United Artists in order to own their own studios and control distribution.)

However, Marx felt certain that not all the contradictions which developed within capitalism were negative. The proletariat may have been dependent upon their wages for subsistence, and been unable to accumulate savings or capital of their own. (This, for Marx, was the definition of the proletariat.) But in the factories of capitalism this perennially exploited class was developing into a skilled and disciplined labour force. This class would have a vital role to play in the next inevitable stage of historical dialecticism. When capitalism collapsed as a result of its own internal contradictions, there would be a revolution and the proletariat would take over the means of production. A 'dictatorship of the proletariat' would then be established. Marx follows this prediction with a characteristically dramatic pronouncement: 'With this social development the pre-history of society ends.'

Yet for Marx the dictatorship of the proletariat was only the first stage. This would be followed by a socialist utopia which very much resembled the woolly vision of Saint-Simon. The struggle between classes, which had been a permanent feature of 'prehistory', would be replaced by a classless society. The state would eventually 'wither away', the old market relations would disappear, money would be abolished, and everyone would receive his or her just deserts. 'From each according to his ability, to each according to his needs.' Apart from these few theoretical pointers, Marx had little to say about the actuality of his socialist utopia. Even Saint-Simon was more specific in his dream.

The dictatorship of the proletariat and the state appropriation of the means of production would lead in the twentieth century to a very different form of dictatorship from the one which Marx had envisaged. Far from withering away, the state expanded into an all-powerful monster – unchecked by competition or opposition. What Marx had not realized was that capitalism's inner contradictions would play a large part in prompting it to evolve, rather than destroying it. Marx was not the only one to misjudge capitalism. None of the great thinkers of his era – from Mill to Nietzsche – had an inkling that capitalism would mushroom in quite the way that it has. What Marx saw as the death throes of capitalism would turn out in reality to be little more than its birth pangs.

Karl Marx finally died at the age of sixty-four in 1883. A dozen friends and fellow-believers gathered at the graveside on that cold March morning in Highgate cemetery. They listened as Engels delivered what must have appeared a hopelessly over-blown funeral oration: 'His name and work will endure through the ages . . .'

Less than seventy years later a third of the world would claim to be run according to his ideas. This venture has now failed, but the force of its beliefs should not be forgotten. Marx's ideas offered the prospect of 'justice on this earth' to countless numbers who had never dreamt that such a thing might one day come. Quasi-Marxist ideas would be espoused, at least momentarily, by such twentieth-century luminaries as Einstein, Bertrand Russell and Wittgenstein, Tolstoy, Gandhi and Nelson Mandela. Many economists now claim that Marx's ideas are of little contemporary relevance. In detail, his critique is said to apply only to the mid-nineteenth-century economy which he analysed (and even this not always correctly). But the larger picture has changed, and continues to change, in line with his key contention. Economics is not an island unto itself; it takes place within society. Who is the economy for? How can its benefits best be shared in a just manner? Such questions which Marx pressed upon economics remain very much alive. We are beginning a century where the division between the First and the Third Worlds ever deepens, where even in the First World the division between rich and poor grows disruptively wide. And not only wealth will need to be rationed more appropriately, in a world whose resources themselves may well be approaching their limits. Marx sought to control the market; the free market has only survived, and will only continue to do so, because it has not only learnt how to evolve, but also how to control itself. This is why economics, for all its flaws and pretensions, becomes increasingly vital to our survival. The people who brought Seattle to a standstill during the first World Trade Organization meeting of the new millennium were not for the most part Marxist extremists or the American poor. No matter their conduct, or the inchoate ideas they expressed: what drove them was a sense of injustice. Others out in the world, powerless and less fortunate than themselves, were getting a raw deal. As Marx in his own way showed, we ignore this fact at our peril.

9

Measure for Measure

Meanwhile another equally revolutionary development was taking place in nineteenth-century economics. Unlike Marxism, this unfolded within the bounds of the classical tradition, the mainstream of economic thought which continued to develop from Adam Smith, through Ricardo and Mill. This innovation was the introduction of mathematics into the central dynamics of economic thought. The aim was to discover mathematical 'laws' which would render economics as certain a science as Newtonian physics.

The statistical method developed in the seventeenth century by Graunt had soon become allied to the relevant concepts of probability theory developed by De Moivre and his successors. A significant step forward in this field was now made by the German prodigy Gauss, generally reckoned to have been one of the three greatest mathematicians of all time (the others being Archimedes and Newton). Karl Friedrich Gauss was born in Brunswick in 1777. His father was of peasant stock, and his mother was illiterate. Myths of Gauss the infant prodigy survive, but the first concrete evidence of his genius dates from when he was a seven-year-old schoolboy. His class of 100 unruly ruffians from the backstreets of Brunswick was controlled by an indolent, brutish schoolmaster named J. G. Büttner. He would pass between the desks wielding his cane, which he used at the slightest opportunity with unflagging enthusiasm. Contrary to conservative belief, this educational method proved so ineffective that in their fear the pupils sometimes even forgot their own names. One afternoon, doubtless exhausted by the rigours of the morning's teaching, Büttner gave the class a long and laborious sum which was sure to keep them occupied for some time, as well as producing a good crop of floggable answers. The pupils were set the mathematically futile task of adding up all the numbers between 1 and 100. The heads bent down, the chalks painstakingly squealing over the slates – except for Gauss, who frowned for a moment, and then simply wrote down an answer. And to make matters worse, it was the correct answer: 5,050. Büttner inquired incredulously how he had arrived at the answer so quickly. Gauss explained that he had simply imagined two sequences of 1 to 100, one ascending, one descending; and then added them up:

$$
\begin{array}{ccccccc}
1 & 2 & 3 & \ldots & 98 & 99 & 100 \\
100 & 99 & 98 & \ldots & 3 & 2 & 1 \\
\hline
101 & 101 & 101 & \ldots & 101 & 101 & 101
\end{array}
$$

The sum of these two sets of 1–100 was thus 101 × 100, a simple sum which came to 10,100. Halve this, and you had the answer: 5,050.

Such mental arithmetic might have been simple enough for a prodigy, yet the initial mathematical inspiration indicated a unique seven-year-old mind. Even Büttner was impressed and immediately made it his task to supply his pupil with the most advanced mathematical textbooks he could obtain. By the time Gauss was eighteen he was making major mathematical discoveries. One of these was the method of least squares, which extended the use of De Moivre's normal distribution and the bell curve. In observations, errors are unavoidable; but Gauss's method sought to minimize the effect of such errors on the final figures. In fact, he discovered that the probable accuracy of the average figure increases in proportion with the square root of the number of observations.

Gauss's first public application of the method of least squares proved sensational. On 1 January 1801 the Italian astronomer Giuseppe Piazzi discovered a new planet, Ceres, now known to be the largest of the asteroids. Unfortunately, after a few days

Piazzi was unable to continue with his observations and 'lost' the planet. For months, astronomers searched in vain for the elusive speck of light emitted by the tiny distant object in the star-filled night sky. Then Gauss decided to apply his method of least squares to the locations at which Piazzi had observed Ceres. As a result, he was able to predict with great acuracy the orbit of Ceres, which was then observed once more, precisely one year after its initial discovery, on 1 January 1802.

Gauss was a complex, reclusive character. When he died in 1855 it was found that he had left many notebooks filled with unpublished mathematical discoveries. A large number of these were written down in his own inconsistent code. For example:

$$\text{Num} = \Delta + \Delta + \Delta$$

This means that any positive number can be written as the sum of three triangular numbers – such as 0, 1, 3 (which can be represented in triangular form as ∴), 6 (ditto: ∴∴), 10 (a further row of four dots below the preceeding triangle for 6), and so forth. Continuing decipherment of these notebooks has revealed that Gauss made several mathematical discoveries which have only recently been 'rediscovered' by mathematicians. Some of his coded formulae have resisted all attempts at decipherment, and may well contain mathematical knowledge of which we still have no inkling.

It would be several decades before the relevance of Gauss's method of least squares to economics was realized. It then transformed the accuracy of economic statistics, both in collection and application, throughout the second half of the nineteenth century. Not until the end of the century did its drawback become apparent. Gauss's method encourages the belief that deviations from the normal distribution are purely due to imprecise observation or a lack of observations. However, exceptions to the norm are not always so simply explained. Rather than dismiss them, it sometimes proves fruitful to investigate them more fully.

The other major mathematical innovations to be applied to

economics in the late nineteenth century were the work of Francis Edgeworth. Born into a wealthy Anglo-Irish landowning family in 1845, he eventually became a professor of economics at Oxford. By all accounts he was a somewhat self-absorbed lecturer. According to one of his students, 'when, after many hours . . . he at last made the supply curve intersect the demand curve . . . one knew it was a great moment. He wagged his beard and uttered inaudible things into it. He seemed to be in a kind of ecstasy.' Edgeworth was pathologically shy, an affliction which resulted in him remaining an unwilling bachelor. His only serious attempt to remedy this dilemma appears to have been a desperate effort in 1889, when a 23-year-old woman recorded in her diary that she was 'being half-heartedly courted by a middle-aged economist named Edgeworth'. The diarist was the future children's writer Beatrix Potter, who decided to reject her awkward suitor in favour of life with the Flopsy Bunnies and Jemima Puddle-Duck.

Following in the footsteps of Adam Smith, Edgeworth's first major work was on ethics. In this, he sought to mathematize the subject using the calculus of variables. After its own fashion, this resurrected the seventeenth-century German philosopher Leibniz's dream that one day all moral disputes would be solved on calculating machines. The watertight logic of mathematics would be used to decide what was right and what was wrong. A little more realistically, Edgeworth based his calculations upon the Utilitarian principle espoused by Bentham and Mill: good is that which brings the most pleasure to the most people. Edgeworth saw human beings as 'pleasure machines', who would always be directed towards achieving the highest amount of pleasure. Here, we can see another attempt at Bentham's 'felicific calculus'. Bentham had developed the concept of *cardinal utility*, which ascribed a measurable quantity of pleasure, or utility, to any particular good. To give a modern example, all other things being equal, the purchase of a modest Honda saloon car might yield 20 units of pleasure to a consumer, whereas for a similar price the purchase of a Honda 1,000cc motorbike might yield 100 units of pleasure. This particular consumer's choice is thus obvious. But once again this comes up against the difficulty of quantifying 'units of pleasure', of measuring emotions. Edgeworth saw a way around this difficulty by introducing the concept of *ordinal utility*. Instead of assigning each good a measurable utility in 'units of pleasure', the consumer was simply asked to list the goods in order of preference. At a stroke, ordinal utility not only produced a realistic list of relative preferences, but also made it potentially possible to link all commodities in a single economy.

The outspoken philosopher and psychic Henry Sidgwick criticized this approach on commonsense grounds. He pointed out that he didn't eat his dinner because he added up the pleasure he would get from it, but simply because he was hungry. This indicated a very real difficulty. There is a categorical difference between need and pleasure – as well as a wide range of physical and mental desire in between. An entire psychology is subsumed in this range of feelings. To reduce such subtleties to a simple scale of utility begs many questions concerning human behaviour. Such economics reduces humanity to an economic cipher: a dangerous circularity.

On the other hand, as Edgeworth pointed out: economics deals with quantities, which means it must be susceptible to measurement, and thus to mathematics. This may be considered as the founding insight of econometrics, that branch of economics where mathematical and statistical methods are applied to economic data. (Though it would be another half century before the actual word 'econometrics' gained currency, and gradually began to assume the proportions of the holy grail to theorists such as von Neumann.)

From the outset, Edgeworth had the highest hopes for the application of mathematics to economics. Here was the 'master key' which would unlock many of the seemingly intractable problems faced by economists. Each problem should be posed like a Euclidian proof. It should be framed in terms of self-evident definitions and mathematical axioms. The problem could then be dealt with by means of rigorous demonstration, step by step, leading to a precise conclusion. Mathematics would thus *prove* the laws of economics, which would be logically incontestable and would follow inevitably one from the other. Edgeworth firmly believed that mathematics could reach to 'the little rills of sentiment and the secret springs of motive where every course of action must be originated'. Well over a century later psychology still resists such mathematicization, even by the back door of neuroscience. Yet with equal obduracy Edgeworth's reductionist belief still remains, as a perennial scientific dream.

Edgeworth's major economic work, *Mathematical Psychics* (sic!), appeared in 1881, just two years before the death of Marx. Edgeworth, as much as anyone, was responsible for those dreaded diagrams and graphs which beset economics, rendering so many of its texts either tiresome or incomprehensible. In *Mathematical Psychics* Edgeworth developed several graphs which remain in use to this day. These include the 'indifference curve', where ordinal utility is extended to combinations of goods, each of which provide equal utility to the consumer (who thus remains indifferent where choice between these combinations is concerned). The essence of this idea can also be stated in plain English. For example, given a choice of eggs and bacon, consumer A might remain indifferent as to whether he has two eggs and one rasher of bacon, or two rashers of bacon and one egg. But outside this range he is not indifferent, having his preferences – ranged in ordinal utility. Edgeworth further developed this idea with the 'contract curve'. This essentially links the indifference range of two consumers, showing where they can exchange goods without gain or loss in utility to either party. For example, consumer B might remain indifferent as to whether he has three eggs or three rashers of bacon, but dislikes combinations or greater quantities. If consumers A and B are faced with a plate containing three eggs and three rashers of bacon, one can gain in utility at the expense of the other. However, a plate containing four eggs and two rashers of bacon, or one containing five eggs and one rasher of bacon, would both appear on the 'contract curve' between consumer A and consumer B, because in either case both could gain a similar utility. It is easy to see how the indifference curve and the contract curve can provide a 'master key' to optimal trading – from the individual to the international level.

The first person to apply mathematics to economic theory in a comprehensive fashion was the Frenchman Léon Walras. His thinking took the French tradition of rationality to the extreme, and remained far too theoretical for many tastes. Yet, as Schumpeter later put it, 'economics is a big omnibus which contains many passengers of incommensurable interests and abilities. However, so far as pure theory is concerned, Walras is in my opinion the greatest of all economists.' How much pure theory and practical economics have anything whatsoever to do with each other has remained a matter of heated debate from the very outset. As early as 1874 the Danish Economic Association saw fit to pronounce all mathematical economics 'offically invalid'.

Léon Walras was born in the Normandy town of Évreux in 1834. His father was an educational administrator who relieved the provincial tedium of his work by analysing economic theory. He had high hopes for his son, but young Léon failed the entrance exam to the prestigous École Polytechnique in Paris. Ironically, the subject in which he failed was mathematics. Instead, Léon entered the School of Mines, with the intention that he would become an engineer. But this was the era of Baudelaire, the first poet of the modern urban individual. The romance and squalor of the new gas-lit Paris soon distracted Walras. He cultivated shoulder-length hair in the romantic fashion, and grew a beard. He also abandoned engineering in favour of literature, for which he showed some talent. At the age of twenty-four he published a novel. That summer his father took him on holiday to Provence, and attempted to dissuade Walras from his romantic notions. As the two of them walked alongside a mountain stream, Walras Senior explained how writers were two a penny: genuine romance lay in the future of science. And one of the greatest tasks that remained to be solved in the nineteenth century was the transformation of economics into a science.

Walras Junior was convinced, but his life in Paris seems to have remained fairly bohemian. He lived in a tiny apartment with his mistress, by whom he had two children, and he supported himself with low-paid commercial work. However, his intellectual endeavours were now directed towards political economy, in the full meaning of this term. He was deeply disillusioned by the reactionary political climate which still pervaded Europe after the collapse of the 1848 revolutions, and he became an idealistic socialist. However, when some Saint-Simonians encouraged him to join their movement, he declined. He evidently drew the line at wearing a red beret and having himself tied into a waistcoat which took a communal effort to divest. He informed them that their socialism was 'unscientific'. At one stage Walras joined up with the grandson of the eminent economist Jean-Baptiste Say (of Say's law) to help found a co-operative bank. Walras even became managing director of the bank for three years, until it failed. (This disaster cured his fellow bankrupt of socialism, but not of financial ambition: Say would go on to court bankruptcy on a national scale when he became a rabidly anti-socialist minister of finance.)

Meanwhile Walras continued to write articles on economic matters, but with little success. However, in 1860 he attended a congress on taxation at Lausanne in Switzerland, where he read a paper whose originality appealed to a member of the Swiss Federal Council. They kept in touch, and as a result ten years later Walras was offered the newly founded post of professor of economics at the University of Lausanne. By this stage he was thirty-six years old, and all but penniless. The idealistic young romantic had given way to a lonely embittered character plagued by a succession of imaginary ailments. He was even forced to ask for an advance on his salary so that he could pay for the railway ticket to get him from Paris and Lausanne. Throughout his long years at Lausanne Walras' temperament continued to decline. His mistress died and he married a rich widow, but he suffered from increasing persecution mania and 'nervous exhaustion'. Yet it was during these years that he produced his supreme work.

In 1877 Walras published his *Elements of Pure Economics*, which dealt with a fundamental unanswered question. Adam Smith's free trade had become an article of faith for the ensuing classical economic tradition. But Walras wanted to know: 'How could these economists prove the results of free competition were beneficial and advantageous if they did not know just what these results were?' In order to answer this question, Walras embarked upon an incisive analysis of general economic equilibrium. This is a concept which is central to Walras' thought, and as a result would become central to economic thinking as a whole. Put simply, equilibrium is reached in a market when the quantity of the product which the sellers place on the market is matched by the quantity which the buyers wish to purchase at the prevailing price. In other words, equilibrium takes place when supply meets demand. *General* equilibrium takes place when equilibrium is reached simultaneously on all markets in an economy, and all products are cleared.

Walras himself likened this situation to an auction, where all the producers bring their goods for sale and all the consumers come prepared to buy. One by one the producers inform the auctioneer of the price for their goods. The auctioneer then calls this out, and the buyers bid for quantities of the goods. When there are too many buyers, the auctioneer adjusts the price upwards; when there are too few, he reduces the prices to clear the stock. But like Edgeworth, Walras had understood that all markets are interrelated: if excess money is spent in one market, it means there is less to spend in another market. So as the price for one set of goods is achieved, adjustments have to be made to the earlier prices which have been established by the auction. By means of this adjusting system there is a *tâtonnement* (literally 'groping') towards a state of general equilibrium. Walras succeeded in reducing this repeated process to a series of precise algebraic equations. In this way he managed to achieve the miracle of giving mathematical form to Adam Smith's 'invisible hand'. And in so doing he 'proved' that free trade worked. Or so he thought.

For a start, this mathematical method has to make a number of crucial assumptions if optimum utility is to be achieved. Firstly, the market must be a 'regime of perfectly free competition'. Also, this state of maximum benefit for all concerned requires full employment if every element in the market is to achieve optimum utility. An equally important proviso requires that all income must be spent, otherwise it would be impossible to find a complete interdependence between supply and demand. If, for instance, a large proportion of the buyers suddenly decided to save half their earnings, demand would immediately be reduced. Walras would later manage to overcome potential objections on this score, by assuming that savings were spent by banks on investment. But many of Walras' other assumptions – such as full employment and unrestricted free trading – certainly had no place in the real economic world. He defiantly countered such objections by demanding, 'What physicist would deliberately pick cloudy weather for astronomical observations instead of taking advantage of a cloudless night?'

What Walras had produced was a mathematical formulation of an interdependent complexity analagous to economic reality. He claimed that in his system, 'no blade of grass could move without altering the position of the stars'. His fiendishly difficult mathematics could be reduced to essentially four sets of equations with four unknowns (price of each good, price of production, quantity of good sold, etc.). But Walras' assumption concerning all income being spent meant that a fifth equation had to be introduced. Unfortunately, in such algebra the number of equations must equal the number of unknowns, or the simultaneous equations remain insoluble. This meant that, despite all his impeccable mathematics, Walras was now unable to produce precise answers from his brilliant equations. The unknown quantities remained unknown, except in relation to each other. Walras managed to overcome this depressing difficulty by selecting one particular good, and using this as a benchmark. The prices of all other goods could thus be established relative to this particular chosen item. This meant that even in Walras' purely theoretical economy there were no absolute prices. It was not possible to establish precisely why a barrel of Fondant de Sion wine cost ten Swiss francs, though it was possible to establish why this was ten times more valuable than a globe of Gruyère cheese.

Walras' analysis depended upon another crucial circumstance. Precisely how an economy could reach general equilibrium depended upon how income and property were distributed at the outset. In other words, there could be different 'modes of economic organization'. That is, society could be organized differently, with more (or even less) equal distribution of wealth, income and so forth. In Walras' view, discussion of such matters lay outside the realm of pure economics. This was a matter of 'art', rather than 'science'. Despite this, Walras held his own strong views on social justice. He was far more radical than his counterparts in England – with the exception of Marx. The ideal society, for Walras, would consist of small agrarian freeholdings – much as was found on the hillsides overlooking Lake Geneva around Lausanne. Such a society would have made the ideal state of general equilibrium much more easily approachable. As a result, Walras' social economic policy – his 'art', as distinct from his 'science' – advocated land nationalization as a prelude to equitable redistribution.

General equilibrium was of course an absolute ideal. It could occasionally be achieved in individual markets, but its achievement in an entire economy was simply the aim. An economy could be seen to be moving closer to, or further away from, the ideal of general equilibrium – and adjustments could be made accordingly. This was the value of Walras' analysis. His theoretical model conceptualized the core of the economic process. Microeconomics, which deals with economics at the level of small groups of consumers, firms or products, now had a central concept to show how it worked, or should work. And the same applied to macroeconomics, which dealt with entire economic systems consisting of an aggregate of smaller individual economic units. Here was a core to which all other economic thinking could be related. Yet as Walras constantly reminded his readers: his analysis of economic equilibrium was intended as a tool, a means of investigation. It was not a picture of how things actually were, or in fact were ever likely to be. General equilibrium was intended as an aim, a blueprint of an optimal situation – with his algebraic formulas simply pointing the way, indicating what could be done.

Despite this, Walras' economic picture has several crucial drawbacks. For a start, his mathematics remains immensely difficult. His formulae are on the whole far too complicated to be used in any real situation. Another flaw was not spotted until nearly sixty years later, by none other than von Neumann himself. Under certain circumstances the solutions to Walras' sets of equations yielded prices of value zero, or in some cases even negative prices. In other words, manufacturers would have had to give away their products free, or in some cases even pay customers to take them away! This suggested the possibility that the entire concept of equilibrium had no practical relevance. The latter difficulty was finally overcome by the American Nobel prize-winner Kenneth Arrow, widely regarded as the finest living economic theorist. Arrow succeeded in a mathematical feat which even Walras himself had been unable to achieve: he managed to prove that a state of general equilibrium could actually exist in an economy. It was not just some theoretical ideal, as had been previously supposed, it could really happen. As a result theoretical economists have returned to pure theory with renewed vigour, producing what the distinguished twentieth-century economist Joan Robinson referred to as 'thickets of algebra'.

In 1892, at the age of fifty-eight, Walras' 'nervous exhaustion' forced him to take early retirement. For the remaining eighteen years of his life he lived in the small town of Clarens, overlooking Lake Geneva. He continued to work on pure economics, but became increasingly distracted by his hobbies of fishing and coin-collecting – apt metaphorical pursuits for an economist.

Walras was succeeded as professor of economics at Lausanne by Pareto, who was to play a major role in extending the scope of mathematical economics. Vilfredo Pareto was born in 1848 in Paris of Italian parentage. His father, an Italian patriot, had fled into exile for political reasons. The struggle had begun which would soon result in Italian unification. Meanwhile, north of the Alps, Prussian influence was extending through the German states. As urban industrialization began to spread, Europe was beginning to coalesce into large nation states.

After taking a degree in engineering at Turin, Pareto travelled Europe for many years as a civil engineer, with occasional ventures into business. At one stage he lost all his savings on the London metal market. His extensive travels enabled him to observe a number of different economies at first hand, especially the advanced British economy and its increasingly successful German competitor. During long, insomniac nights in lonely provincial hotels he avidly devoured all the works he could find on political economy. Finally, at the age of forty-one, he married a contessa of Russian extraction. By now he had published a number of original economic papers. His exceptional ability to translate economic rhetoric into precise mathematical formulae led Walras to recommend him as his successor at Lausanne in 1893. Six years later Pareto inherited a fortune of 2 million gold lire from an uncle, whereupon his wife ran off with one of their new young servants. Pareto consoled himself by moving into a luxurious rural residence overlooking Lake Geneva, which he named Villa Angora. Here he lived in solitary splendour off the finest vintage wines, dining amongst his ever-increasing menagerie of thoroughbred angora cats, whom he regarded as examples to humanity. At the same time he carried on a vituperative polemic against all manifestations of sexual puritanism. By now he had come to dismiss mathematical economics as hopelessly narrow, and utterly unsuitable for application to human behaviour – which was only 'make-believe logical'. Instead, he discovered the 'logic of the senses': this was the illogical logic which was responsible for all human action and the whimsies of human emotion. These non-logical acts he called 'residues', which were determined by two elements. They were both an expression of underlying human psychology and an indisputable fact which is incapable of further explanation.

It remains unclear whether Pareto's despair at human behaviour was inspired by introspection or feline activities (angoran or spousal). Fortunately, irrational self-contradictory aspects did not interfere with his exceptional analytical abilities. A glance at any economic dictionary will reveal a number of lasting economic concepts bearing his name. The notion of 'Pareto Optimality' is clearly developed from Walrasian general equilibrium. Starting from basic principles, Pareto noted that two people will only trade with each other freely if both of them gain something out of the exchange. If one stands to gain and the other stands to lose, trade will not take place. Pareto Optimality is reached when an overall economic situation is such (as, say, in general economic equilibrium) that it is impossible to make one person better off without making another worse off. Pareto Optimality describes a situation of complete economic efficiency: no overall improvement is possible. It has also been translated into social terms, regarding the optimum use of a society's resources. This interpretation provides the compelling intellectual argument for modern welfare economics.

Pareto's most far-reaching and controversial mathematical application resulted in his law of income distribution. Making a widespread study of the relevant statistics in America and countries throughout Europe, Pareto argued that there was a universal pat-

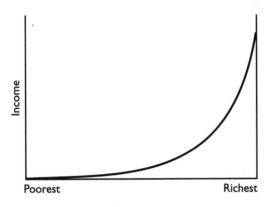

Pareto's income distribution curve

tern of income distribution. Astonishingly, this appeared to be the same in all societies, and to have applied throughout history. Previously it had been believed that the distribution of income would vary, depending upon the economic structure of a society. Pareto showed that this simply was not true. In both agrarian Russia and industrial Britain the distribution of income was the same. Similarly, as far as it was possible to determine, this distribution had followed the same pattern in ancient Rome as it did in nineteenth-century Britain. The actual pattern of this distribution also came as something of a surprise. Pareto showed that contrary to popular expectation or belief, incomes within a society did not rise proportionately in a straight line. Instead, they rose geometrically. Income rises by a minimal amount amongst the majority, the lowest-paid families, but shoots up amongst the wealthy few. According to Pareto, this pattern directly mirrored the distribution of intellect, talent and enterprise within a society. Very few deserved to be rich, most deserved to be poor. Pareto's figures are for the most part undeniable. His explanation of them is not. This distribution of income is almost inevitable with unrestrained free trade. Under such circumstances, those who have wealth have sufficient power and political influence to thwart any attempt at a more equitable redistribution. This is perhaps best illustrated by attitudes towards income tax during the period. In Britain, income tax was not accepted as a permanent levy until the 1880s. In the following decade income tax was declared *unconstitutional* in the United States. Why should the rich pay for the poor? This state of affairs was not remedied until 1913. By then an answer had been found as to why the rich should pay. They benefited more from society, and should therefore contribute more towards it. The obvious contemporary parallel is with the ludicrous remuneration given to bosses of large corporations. Despite all reasonable arguments to the contrary, these few individuals continue to exert sufficient power for their pay structures to remain free from any interference – either by the legislature or by shareholders. This too would be unconstitutional.

Pareto's discovery concerning income distribution makes unpalatable reading for the overwhelming majority of us who are not members of the filthy rich. It raises a number of uncomfortable questions about the very nature of human society, and how it evolves. Is such economic and political ordering inevitable? Pareto believed that this income distribution inevitably solidified into a class structure, as those of ability sought privileged confirmation of their superiority. Or, in other words, those who had got their hands on the lucre wished to implement a structure by means of which they could both keep it and display it with impunity. However, Pareto was the first to admit that members of classes rose and fell. This led to a 'circulation of elites'; history was 'the graveyard of the aristocracy'.

Pareto's political opinions reflected his view of human behaviour. Irrationally, he believed in republicanism but not democracy. The latter was a luxury which society could ill afford. Liberty, like rationality, was after all an illusion. Such ideas were seen as presaging the era of fascism, and were welcomed in Italy when Mussolini took power in 1922, the year before Pareto's death. However, Pareto's 'pessimistic realism' was related to mathematical fact rather than master-race fantasy.

Pareto's income distribution curve was a fact, however unpalatable. And it remains one. Yet its worst effects have arguably been ameliorated over time. In the First World economies a more equal distribution of income has not only been achieved, but has been shown to work for the benefit of society as a whole. And ironically, it is the notion of Pareto Optimality which has contributed to this. An economy works better when everyone benefits. Despite such developments, there is no denying that Pareto's income distribution curve continues to cast its shadow. Even First World economies still have their sizeable underclass; and a fat cat corporation boss can receive in a month more than many of his workers will spend in their lifetime. Meanwhile in less advanced economies, such as Brazil and India, Pareto's income distribution curve survives as if carved in stone.

Walras well understood that the truths of his pure mathematical economics were not reality. Pareto chose to have it both ways. His optimality is pure mathematics, his income distribution is derived from less pure statistics – a field which deals intimately with reality, yet seems to encourage practitioners to be economical with its truths.

The English economist Jevons, on the other hand, was convinced that his mathematical economics was not only true but indistinguishable from reality. William Jevons was born in 1835 in Liverpool, where his father was a prosperous iron merchant. Jevons Senior was making a fortune out of the railway boom, and also built one of the first iron ships to remain afloat (at least temporarily). For young William the pleasures of a privileged upbringing, complete with private tutor, were blighted by familial tragedy. His mother died when he was ten, and two years later his older brother went mad. Then the railway boom ended, and in 1848 Jevons

Senior went bankrupt. Young William managed to gain a place at University College London, where he showed promise at chemistry. But he ran out of money in his second year, whereupon he took the drastic step of setting sail for Australia to become assayer at the Sydney Mint.

Jevons arrived to find the Australian gold rush in full swing. This was the era of the great gold rushes, which had started with the famous '49ers in San Francisco. Urbanization and emigration to America and the European colonies had resulted in a new rootlessness. The European colonial powers had by now laid claim to almost all the large unexplored tracts of the globe and were seeking to extend their influence still further. In mid century the British and the French fought the Opium Wars against China, consequently establishing trading stations along the coast and forcing the Chinese to buy imported opium. Economic theory was making great advances in Europe; meanwhile its practical enterprises overseas were developing new sales techniques (gunboats, opium pushing) and ceaselessly seeking new sources of raw materials (gold rushes, poppy fields, etc.).

The Australian gold rush of the 1850s saw mobs of prospectors from Ireland to China, ranging from the sons of aristocrats to naval deserters and snake-oil salesmen, arriving off every boat. An indication of the scale of this 'rush' is shown by the population figures for the state of Victoria, its epicentre. During the 1850s these rose from less than 80,000 to over a quarter of a million. In their wisdom, the British government took the drastic step of stopping the transportation of convicts to Australia in 1853. There seemed little point in spending taxpayers' money shipping criminals to a land full of gold. Meanwhile large consignments of gold were brought to Sydney and kept under armed guard at the Mint, where Jevons dutifully assayed minute samples for purity in his laboratory. (A few decades later the Sydney Mint would take its place in economic history as perhaps the first institution of its kind to produce official forgeries. These were Austrian Marie-Theresa silver dollars, for use by British traders in the Middle East and Africa, where such coins were the only trusted currency. According to numismatic lore, these coins – all minted from the same 1780 die – only ceased to be used around forty years ago. However, I know from my own experience that they remained currency at remote markets in the horn of Africa until well into the 1990s.)

Jevons fulfilled his duties at the Sydney Mint for the next five years. By his own admission he 'made no friends in Australia'. But his time was not wasted. He began reading Adam Smith's *Wealth of Nations*, which he found 'an excellent though rather old book'. This set him to thinking about political economy, and he read several other books on the subject. (One chauvinist British commentator has cast doubt on this fact, claiming that there would not have *been* any other books on this subject in Australia at this time.) Be this as it may, Jevons's pondering upon economic theory evidently had its effect, for he soon came to the conclusion that he possessed an exceptional mind capable of highly original thought. Nothing unusual about this in a 22-year-old – apart from the highly unusual fact that it was true. Jevons was soon producing original ideas on such wide-ranging topics as the gold standard and the falsity of inductive reasoning.

In 1859 Jevons sailed back to England, where he returned to University College, this time to study economics. The course was largely devoted to Ricardo and Mill, which he found 'an extraordinary tissue of self-contradictions'. Fortunately, he also attended lectures by the formidable one-eyed mathematician Augustus De Morgan, who was developing an entirely new form of logic. After graduating, Jevons continued work on De Morgan's 'combinatorial logic', which reduced the laws of reason to a number of general formulae. These involved basic terms or classes which could be substituted for each other. Jevons quickly grasped the principle of this, turning these basic laws into simplified algebraic form. Here was the beginning of modern symbolic logic. This new algebra proved particularly amenable to mechanical manipulation, and Jevons even constructed what he called a 'logical machine', which he showed to the Royal Society. This is now recognized as a significant forerunner of the modern computer, incorporating several of its basic principles.

Yet all this mathematics was for a purpose. 'There can be no doubt that pleasure, pain, labour, utility, value, wealth, money, capital, etc. are all notions admitting of quantity; nay, the whole of our actions in industry and trade certainly depend upon comparing quantities of advantage and disadvantage.' Once everything was expressed in mathematical terms, the woolly concepts which had passed for economic thought in the past would simply be rendered redundant. Where Walras concentrated on equilibrium, Jevons focussed on value, or utility. For him the central economic factor was the price of goods in the market-place. Previously Marx, and to a more or less extent most of the classical theorists, had seen value in terms of the cost of production. Jevons took the opposite approach, which in practice turned out to be more realistic. Things acquire their value in the market-place, not in the factory. And what matters most in the market-place? Ultimately the consumer, not the price of production. Jevons saw value in terms of consumer satisfaction. In other words, he proposed a theory of value in terms of utility – the satisfaction or fulfilment of need produced in the consumer. This was what determined the price the consumer was willing to pay for a product. However, Jevons also recognized the concept of marginal utility. 'As the quantity of any commodity, for instance, plain food, which a man has to consume, increases, so the utility or benefit derived from the last portion used decreases in degree.' In other words, the *margin* of utility decreases as consumption increases. Consumption of one ice cream by a customer produces a certain utility, consumption of two by the same customer produces proportionately less, of three even less, and so forth. Marginal utility is a vital factor in quantifying the satisfaction of the consumer, the demand for goods on the market, and ultimately the price of a product: it is vital to value theory. Jevons was the first to introduce consumer behaviour as a major factor in economic analysis. Unlike Walras, he took account of mathematics *and* psychology – hence his insistence that his mathematics *does* apply to reality.

But does it? Edgeworth suggested that we can establish relative 'units of utility', and thus an exchange theory of value which embraces the entire market. But this remains one-dimensional, in the light of marginal utility, which takes account of a consumer purchasing more than one of the same item. Yet the question remains: does marginal utility apply to the entire market, to all consumers? The answer is of course no. Some people will get more satisfaction from the second ice cream, or the second (or third or sixth) pint of beer. Consumers are not always sensible, or even rational. It may not be sensible to spend all your money on beer on Saturday night – but many do, and reckon the experience well worth it. The average individual in an economic equation can bear little relation to the recalcitrant personalities we encounter in everyday life (or even in the mirror). Like Walras, Jevons represents one step forward, and one step back. An immense advance in theory, yet another retreat from reality.

Jevons ended up as professor of political economy at University College London. Other academics were the first to appreciate the subtlety of his analysis, and would attempt to build upon it. This is what economics was now in danger of becoming: simply an academic pursuit. Meanwhile practising economists continued trading in 'genuine forged' Marie-Theresas, and getting the Chinese hooked on opium. (Indeed the opium trade presents something of a paradox for marginal utility analysis: decreasing satisfaction from increasing amounts, yet increasing need for increasing amounts.)

Jevons himself was an introverted, very private creature. Though he married and had children, his family found him difficult to live with. His personal journal indicates a manic-depressive temperament, which found solace in the protracted melancholy ecstasies of Wagnerian opera. His solitary thinking was not without its eccentricities. At one point, he produced a highly ingenious explanation of the business cycle, the regular wave-like fluctuation between boom and recession exhibited by an economy as a whole. Thinking on such matters had remained somewhat sketchy. Malthus had produced a perceptive outline of 'gluts' (depressions); meanwhile even a thinker of Ricardo's stature had simply denied that they existed. When the British economy fell into the depression that would last throughout the 1880s, Jevons decided that it was time to apply mathematics to this phenomenon. After studying the figures for the previous 150 years he calculated that business cycles lasted on average 10.46 years. This figure rang a bell. Jevons had read somewhere that sunspot activity occurred on average every 10.45 years. Such similarity was too close to be a mere coincidence. Evidently there was a link between economic activity on earth and vortexes of flaming gas on the solar surface. Jevons's colleagues did not welcome this attempt to raise economics into the realm of astrology. Nevertheless, Jevons was one of the first to attempt a (quasi-)serious analysis of the business cycle.

Jevons also applied his considerable expertise in mathematical analysis to the coal industry, the driving power behind Britain's imperial supremacy. He pointed out that, if Britain continued to consume coal at its current rate, its price would soon rise and put the country out of business. And worse would follow, as the world's major energy resource finally ran out altogether. Even the most strenuous conservation efforts could only postpone this calamitous day: the end was inevitable. When Jevons published *The Coal Question* in 1865, it immediately made him a household name. He took over the Malthus role as prophet of doom amongst the chattering classes. Meanwhile natural gas had already been in commercial use since 1821; electricity generators had been on sale in most European countries since the 1850s; and by the 1860s oil was being drilled in half a dozen US states. Hindsight is easy –

but then so is doom-mongering, an industry that shows no sign of running out of energy at the start of the twenty-first century. Jevons was no simple self-serving Jeremiah: the prospect of scarcity evidently struck some psychological chord. At one point he became convinced that supplies of paper would run out. Whereupon he bought so much that his family was still using Jevons's hoard *fifty years* after his death. This came about prematurely at the age of forty-six, as a result of a swimming accident off the Devon coast. At the time he was unwell, both mentally and physically, and the swim was against doctor's orders. In true Victorian fashion, his death was put down to a heart attack, but the suspicion of suicide remains.

10

Into the Modern Age

The classical economic tradition of Smith, Ricardo and Mill had taken a mathematical turn. But the development of economics was not all mathematics and moonshine (or sunspots). A transformation of the old classical tradition would now take place. This would become known as neo-classical economics. Its founding father is generally seen as the Englishman Alfred Marshall, not least because he wrote the textbook from which neo-classical economics would be taught for generations to come.

Alfred Marshall was born in 1842, the son of a clerk at the Bank of England and a butcher's daughter. By all accounts, Marshall Senior was an even worse ogre than John Stuart Mill's father, regularly interrogating his son until eleven at night on the intricacies of Hebrew grammar. Young Alfred's summer holidays on his uncle's farm, where he developed an enthusiasm for shooting dumb animals, appear to have kept him sane. At school, in the words of one of his masters, 'he had a genius for mathematics'. Marshall's father neither approved of, nor understood, mathematics (an unusual approach for a ledger-clerk at a central bank). Instead, the precocious young Alfred won a scholarship to Oxford to study classics, his father's intention being that he should enter the Church. But in Keynes's lyrical words, 'No! he would not be buried at Oxford under dead languages; he would run away – to be a cabin-boy at Cambridge and climb the rigging of geometry and spy out the heavens.' Fortunately, Marshall managed to persuade his uncle, who had made some money out of the Australian gold rush, to pay his fees, and supported himself by coaching fellow students in mathematics.

At Cambridge Marshall came under the influence of the philosopher Sidgwick, who had been such a thorn in Edgeworth's side. It was said that Sidgwick was possessed of every Christian virtue, and even believed in the afterworld – with whose inhabitants he remained in psychic communication. Yet he drew the line at believing in God, and remained a convinced atheist. This paradoxical attitude must have had a paternal ring for Marshall, and Sidgwick soon became a father figure to him. Marshall too became a firm believer in all Christian virtues, except belief – a faithless faith he would retain to the end of his days. Besides philosophy, he also began studying maths – but in secret. (God the father may not have existed, but his wrath lived on in London.) In philosophy Marshall became more and more intrigued by the problems of ethics. One day, in the midst of a typically passionate undergraduate argument, he was chided: 'Ah, if you understood political economy you would not say that.' Marshall immediately set to reading Mill. Unlike Jevons, Marshall 'got much excited' by what he read of Mill and soon began seeing the problems of ethics in terms of economics. 'In my vacations I visited the poorest quarters of several cities and walked through one street after another, looking at the faces of the poorest people.' In the slums of London's East End he was particularly struck by the amount of 'cringeing wretches'.

Pareto's income distribution curve was more than just a line in a diagram: it represented widespread human misery. Several decades had passed since the Manchester Engels had described, but conditions amongst the poor had improved little. With the death of Dickens in 1870, London's populous East End largely disappeared from the view of so-called respectable society. In more ways than one: according to meteorological records, the sun would sometimes only be visible through the choking dirty yellow fog for a total of five hours throughout the *entire month* of January.

(The reason why the slums of many northern European cities tended to be in the eastern districts was because the prevailing westerly winds blew the smoke from all the domestic chimneys and belching factory stacks in this direction.) Only the occasional foreign writer penetrated the murky, teeming backstreets of London's East End: in the 1870s the French poets Rimbaud and Verlaine in search of the opium dens of Chinese Limehouse; Joseph Conrad in the 1880s on his way to the East India docks to join a ship; Jack London dressed as a tramp with a gold sovereign sewn inside his jacket just in case. It took the exploits of Jack the Ripper during 1888 to bring this 'sluggish tide of general misery' momentarily to the vicarious notice of Victorian society.

As a result of what he had seen, Marshall decided to devote his life to economics, which he defined as 'the study of man in the ordinary business of life'. Economics as we know it today begins with Marshall – for better or worse. It was he who popularized the use of all those 'simple explanatory diagrams' which make economics so blindingly clear, as well as many basic concepts so brilliant that they frequently remain beyond the comprehension of those who need to understand them. However, his major contributions overcome both these drawbacks.

This is most tellingly illustrated by Marshall's classic supply and demand diagram. The rising curve for supply illustrates the law of supply: as the price becomes higher, so firms will produce a greater quantity of goods. The falling demand curve illustrates the law of demand: as the price falls, so consumers will purchase a greater quantity of goods. Equilibrium is reached in the market where supply and demand meet. At a stroke, Marshall solved the conflict between the two competing theories of value – the classical theory, which saw value as the cost of production, and Jevons, who saw price determined by marginal utility, the satisfaction of the customer. According to Marshall, the value (price) of a good was determined by 'the scissors of supply and demand' – as illustrated in figure 3.

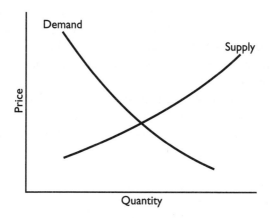

Marshall's supply and demand diagram

Marshall also greatly extended the idea of equilibrium analysis. Walras had conducted general equilibrium analysis, which involved an entire economy (or macroeconomy). Marshall introduced the notion of *partial* equilibrium analysis, which concentrated on a particular sector of an economy. This could be limited to a particular industry (e.g. the motorbike industry), a particular firm (e.g. Harley-Davidson) or even a single individual (e.g. a particular owner of a Harley-Davidson). Partial equilibrium analysis enabled a much more detailed study of the different factors affecting the economic sector under study – such as mushrooming growth in the Third World small motorbike market; whether Harley-Davidson should consolidate its position in the luxury bike market, or expand into the lower range; high-hire purchase payments pushing the individual motorbike owner into debt.

Walras had been forced to make a number of big assumptions (such as full employment, and all income being spent); Marshall was forced to make even more. The most important of these was the proviso 'other things being equal'. That is, the assumption that factors outside the economy under study would not seriously affect its inner workings. There would be no 'feedback effects' on the motorbike industry from a slump in the Far East; from Harley-Davidson being the victim of a hostile takeover bid; or from 'downsizing' putting the individual motorbike owner out of a job. 'Other things being equal' represented a major assumption, but the effect was lessened when the economy under study was small in comparison to the economy as a whole. Thus, the economist could come to grips with the host of factors which might bring about the optimum state of equilibrium, the most efficient possible economic state of affairs for that sector. These might include motorbike manufacturers switching assembly to countries with cheap labour costs; Harley-Davidson deciding to lend their name to a range of men's luxury products rather than go downmarket; the individual motorbike owner's decision to sell his Harley-Davidson and buy a cheaper bike.

Investigation of this multiplicity of factors led Marshall to come up with the important concept of 'elasticity'. The scientific view of economics attempts to relate all factors in an overall scheme of cause and effect. Elasticity measures the size of this causality. A tiny rise in one variable can have a huge effect on others: a relationship which is said to be elastic. For instance, a comparatively small rise in the price of houses will have a highly elastic effect throughout the economy as intending house-buyers adjust their expenditure accordingly. When the effect is small, the relation of cause and effect between one variable and another is said to be inelastic. Marshall noted that price was a big factor here. For goods with a low price (such as pencils or beer) a comparatively large percentage rise will have little effect on the price of other goods. This is why finance ministers have no hesitation in slapping extra tax on beer in the budget, yet increase taxes involved in house purchase at their peril.

Perhaps the most important factor which Marshall introduced into economics was time. Previous considerations of the subject had tended to include insights, concepts or calculations which dealt with an essentially static situation. Admittedly, Adam Smith's 'invisible hand' certainly moved in time, but it didn't concentrate on the *effect* of time. Even Walras' equilibrium was essentially a static notion, as its name implies. According to Keynes, Marshall saw the overall picture as 'a whole Copernican system, by which all the elements of the economic universe are kept in their places by mutual counterpoise and interaction'. Interactions and adjustments take place in time, *and the time they take will have its effect*. In the short term, adjustments in price will be affected by supply and demand. The price of laptop computers drops as they flood the market. But in the long term the cost of production will play the major role in determining the price: the cost of assembly and shipment of laptops will ultimately limit the decrease in price. It is time which acts as the counterbalance – between supply/demand and cost of production – in the determination of value. Time will also play a major role in elasticity. Some prices will take a longer time to filter through an economy before they have an effect. A rise in the price of coffee beans takes time before it affects the cost of coffee on the supermarket shelf. Others will have their elastic effect almost at once. The rise in the price of coffee at the supermarket has an immediate effect on sales.

With Marshall, economic analysis began to expand into the very nitty-gritty of commercial life and its processes. This was also true of those who studied this new subject. As students who attended lectures by Marshall and other like-minded economic academics began graduating into the real commercial world, neo-classical ideas were increasingly put into commercial practice. Businessmen and factory owners, politicians and civil servants, even journalists – all quickly saw the benefits of economic analysis. Understanding the ways and processes of the market, and how to act upon them, gave commercial advantage over competitors. But how did this work, and to whose advantage?

The idea of economics had grown to the point where its effect could be seen in every aspect of human activity. Ironically, in doing so, its investigations had become increasingly theoretical. Indicative of this development is Marshall's introduction of time into the economic equation. This time was abstract, not historical or personal time. No events other than economic developments took place in this time. Economics had come a long way from Adam Smith's *Wealth of Nations*, whose discursive scholarship had sought to include everything – from historical developments to moral concerns. How could such things possibly be included in equilibrium analysis? Morality is difficult to measure, yet its exclusion from the economic equation was now becoming a source of increasing concern – to more than just utopian socialists and Marxists.

One of those who raised this question was the little-known eccentric Swede Knut Wicksell. While Marshall was writing text-book economics, Wicksell was producing practical ideas with the aim of 'correcting' neo-classical economics. Just two years after Karl Marx had vacated his seat in the British Museum for ever, Wicksell sat himself down in the very same library and proceeded to teach himself economics. It was 1885, and he was thirty-four years old. By this stage he had already published a book of 'social indignation' poetry, been a devout Christian and become an equally devout atheist, and established a reputation as a sensational speaker on anti-royalism, anti-marriage, anti-celibacy and many other topics bound to stir up controversy in provincial, conservative Sweden. Typical of these was his position as a 'defence nihilist'. This involved the admission that Sweden could never defend itself against a major nation, so there was no point in having an army. The defence budget should be used for social benefit, and sovereignty should be surrendered to Russia. Sweden's long tradition of democracy would then help transform Russia's autocratic rule from within. His speeches on such topics had predictable effects in both the public hall and the press, firmly establishing him as a student hero. He would later go on to live openly with a young Norwegian gym instructress by whom he had two children; serve a short spell in prison for ridiculing the Immaculate Conception; and finally obtain a degree in mathematics after fifteen years of study.

Wicksell's economic ideas were equally ahead of their time, and were to prove prophetic well beyond his native Sweden. Despite keeping Marx's seat warm at the British Museum, Wicksell was to be no Marxist. His views on this were unequivocal. Unlike Sweden's defence policy, capitalism worked. Even neo-classical economics worked, up to a point. It was the neo-classical *economists* who were at fault. Despite dealing with increasingly wide-ranging aspects of economic activity, the neo-classical ideas of Marshall and his followers were taking on an increasingly inhuman aspect. Equilibrium, both general and partial, established optimum market efficiency. The human reality was ignored. Excess profits and excess hardship were not factored into Walras' equations or Marshall's analysis of partial equilibrium. Teaching at Cambridge, Marshall had become cut off from the squalid slums which had prompted his initial compassion. On the shores of Lake Geneva, beneath the snow-capped Alpine peaks, Walras had been a world away from the spewing smoke-stacks of the German Ruhr and England's evocatively named Black Country. As professors, their concern was with the effect of ideas upon ideas, not man's inhumanity to man. Their concern was with the scheme of things, not with those who inhabited this scheme.

Even when Wicksell finally became a lecturer in economics at the Univeristy of Lund in southern Sweden, he never lost sight of what was actually happening in the market-place. Indeed, in a very literal sense, he brought it with him. On the way to lecture, he would stop off at the local market to do his shopping. He would deliver his lectures in a fisherman's cap, with bags of vegetables, packages of fish or meat, and assorted fruits hanging from the lectern. The symbolism, like the reek of herring, may not have been intended – but it was plain for all to see where he had come from.

Untrammelled free competition, even at optimum efficiency, did not in fact produce an optimal distribution of resources. Not for the people involved, at any rate. Some were ground down, others over-rewarded. This surely was inefficient. But the reason why it was inefficient was not the fault of the free-market system. The way Wicksell saw it, capitalism assumed 'that from the beginning all men are equal. If that were so everyone would be equipped with the same working power, the same education and, above all, the same economic assets . . . each person would [then] have only himself to blame if he did not succeed.' It was the business of the state to try and reduce these inequalities – by means of such things as inheritance tax and equality of education. The distribution of income should be according to marginal productivity. In other words, the more an individual contributed to the overall margin of profit, the greater should be his income. The free market forced competing firms to operate at their optimal size. The government should take over any market where competing firms operated a cartel, keeping prices artificially high and ensuring excess profits. It should also take over any market which tended towards a monopoly, such as railways, the postal service and the new electricity suppliers. Wicksell insisted that a course could be steered between the rock of untrammelled free trade and the hard place of state control. Free competition or monopoly were not the only options. Here we see the blueprint for the mixed economy, containing both a private sector and nationalized industries. Despite nationalized industries being sold off worldwide during the last couple of decades, all capitalist economies remain to an extent mixed.

Wicksell also turned his attention to another central concern: money. The idea of money is both obvious and elusive. What is it? What does it do? Does it have an ultimate value (other than itself)? Is it a means or an end? Up to this point neo-classical economists had regarded the ebb and flow of money on the market

246

as a mere 'veil' which obscured what was actually going on – namely, the transfer of goods from seller to buyer. Wicksell understood things differently. Money, and its availability, directly affected the economy. The availability of money now depended to an increasing extent on the banks. More and more enterprises, both large and small, relied upon a bank loan to get started. The rate of interest at which this money was loaned was thus all-important. A high rate of interest meant less would, or could, borrow; a low rate encouraged entrepreneurial risk.

Wicksell identified two distinct interest rates. The 'market rate' was that charged by the banks. The businessman borrowed money from the bank at the market rate in order to set up a new factory, invest in new machinery or take on more workers. The 'natural rate' was the rate of return he made on the money he had invested in the factory, machinery or increased manpower. If the natural rate exceeded the market rate, he would profit accordingly. The gains from the investment would be higher than the cost of repaying the loan. However, Wicksell was able to show that if the natural rate stays above the market rate, this will cause an uncontrolled expansion of the market accompanied by an increase in prices. On the other hand, if the natural rate falls below the market rate, investment will fall and unemployment will rise. According to Wicksell's analysis, the market rate should be kept as close as possible to the natural rate, and this could only be done by the government or central bank fixing this rate. There was a demonstrable need for such a monetary authority. Here was an example of the government, or central bank, intervening to ensure the smooth running of the economy – which would falter if left to its own 'free' devices. Once again, there was a cast-iron case for a measure of government control of the economy. Central banks increasingly began to recognize this. Indeed, it is why today, in the freest economy in the world, the US Federal Reserve (most famously in the form of Alan Greenspan) continues to fix the interest rates.

Even in the United States, government intervention in the economy has a long and decisive tradition. By the latter part of the nineteenth century the United States was in the process of becoming the largest economy in the world. The European powers, whose empires now occupied over three-quarters of the known globe, remained for the most part blissfully unaware of the threat from this Johnny-come-lately. The modern economist Paul Ormerod has noted the figures, which speak for themselves. In 1870 Britain, with a population of around 30 million, was still the world's leading economy. Meanwhile the United States had an income per head of 80 per cent of the British figure, with a population of nearly 40 million. As Ormerod puts it, 'The combination of a higher population and a lower income per head meant that the size of the American domestic market was virtually identical to that of the British.' By the end of the first decade of the twentieth century the balance had shifted sensationally. By now the United States' income per head had leapt from being 20 per cent behind Britain to 20 per cent ahead. Meanwhile the British population had risen by nearly 15 million, but the United States population had risen by almost 60 million. On equal terms in 1870, the American domestic market was now *250 per cent larger* than that of Britain, its nearest competitor.

During the latter part of the nineteenth century the world economic scene underwent a sea change. The modern world was coming into being. As early as 1856 commercial refrigeration was introduced in America and Australia. Soon ships were carrying frozen lamb halfway round the globe from Australia to Europe. Railways were by now expanding to their limit throughout Europe and America. In 1869 a ceremonial gold pin was driven into a railway sleeper in the Utah desert at Promontory. The Union Pacific railway (from the east) was now linked up with the Central Pacific (from the west coast): America was joined coast to coast. Such transport networks were matched by communications links. The old-fashioned Morse telegraph wires had spread alongside the railway lines, but these were soon superseded by the telephone (patented by Alexander Graham Bell in 1876). By 1887 the United States had over 150,000 telephones. Far from energy running out, as Jevons had predicted in 1865, new sources such as electricity and oil now came into play. In 1891 Paris became the first city to have its streets illuminated by electric light, causing it to be named the City of Light. On the high seas steam was replacing sail. By 1893 Boston was linked to Chicago by telephone. Goods could now be ordered instantly by telephone, transported speedily by rail or steamship, even preserved by refrigeration, and factories could continue production at night by electric light. The era of the mass market had arrived.

Technological advance meant that production costs were becoming ever cheaper. Prices too decreased, though at a some-what slower pace. There was an unforeseen reason for this. Neo-classical models had predicted that developing technology and production costs would limit firms to a comparatively small size. Markets would thus consist of a healthy number of competitors, none of whom would be able to achieve market domination to the point where it could influence prices. According to the neo-classical economists, who were now taking up influential government posts in America as well as Europe, this was how the free market would continue to evolve. But such models had overlooked the fact that evolution can also involve co-operation (even when the rules do not permit this). Markets took to forming 'trade associations', whose covert purpose was to agree to limit output and maintain price levels, thus ensuring increased profits all round at the expense of the unwitting consumer.

In 1890 the US government outlawed such cartels with the Sherman Anti-Trust Act, brought before Congress by the younger brother of the celebrated Civil War general after whom the tank was named. But the Sherman Act did not drive a tank through the cosy cartels as intended. Indeed, it had the very opposite effect. With prices freed up, the larger companies quickly began undercutting

249

their smaller rivals, then took them over. What had previously been unspoken agreement now became market fact. A flurry of mergers and buy-outs created vast new conglomerate companies, which were then able to control prices through their sheer market muscle. Huge profits were soon being made, and as before it was the public who paid. Several of the giants formed in this bonanza remain household names to this day: Union Carbide, Quaker Oats, Heinz, Eastman Kodak, American Telephone and Telegraph (AT&T), to name but a few. Ironically, the only effective use of the Sherman Act during the first decade of its existence was against the embryo trade unions, which were declared illegal cartels conspiring to force up prices and limit production. By the turn of the century only 3 per cent of the US labour force was unionized. Untold wealth was now being generated in America, but it remained in the hands of the few.

Meanwhile in Europe the trade associations used their powerful influence behind the scenes to ensure that no anti-trust legislation was passed. As a result, the smaller European companies were no match for the emerging US giants, who were now laying the foundations for American dominance in world markets. This was the legendary age when financial (and moral) dinosaurs stalked the American market, savaging all around them at will, leaving behind a trail of devastation bearing their name. J. Pierpoint Morgan all but gained control of the US railroad system. Anecdotes of his power still astonish. In 1895 he bought enough gold to keep the US dollar on the gold standard; twelve years later he personally averted a Wall Street crash. (Alarmed that so much financial power had fallen into a single pair of hands, the government set up the Federal Reserve.) The father and son partnership of Meyer and Daniel Guggenheim was running the US mining industry by 1901, with controlling interests worldwide – ranging from Bolivian gold to West African diamonds, from Congo rubber to Alaskan copper. Cornelius Vanderbilt (shipping; 'I will ruin you') and Andrew Carnegie (US steel; 'Watch the costs and the profits will take care of themselves') were in the same league. Such men didn't give a damn for economics. They killed by instinct, leaving accountants to mop up the blood. Money had no philosophy. What ultimately was money? Did it have any value other than itself? These were not questions that concerned them. In their hands money was unstoppable. As J. Pierpoint Morgan put it to a judge, 'I don't know as I want a lawyer to tell me what I cannot do. I hire him to tell me how to do what I want to do.' Money had taken hold in American society. Money could do anything. By 1919 it could even buy the outcome of the baseball World Series – surely one of the defining moments in that unending saga known as America's loss of innocence.

Innocence may have been lost, but economic control was not. Not quite. The parable of John D. Rockefeller, the greatest of all the robber barons, plainly spells this out. In 1863 the 23-year-old Rockefeller set up his first oil refinery amidst the then booming oil fields of Cleveland, Ohio. Two years later he founded Standard Oil, and by 1872 Standard was running all the refineries in Cleveland. Just eight years after that, Standard Oil controlled 95 per cent of all the oil produced in the US. The skulduggery involved in such megalomaniac expansion left a rubble of destroyed hopes in its wake. But in 1892 Rockefeller ('God gave me my money') was prosecuted under the Sherman Anti-Trust Act. For nineteen long years he fought tooth and nail, shifting the company from state to state and using every means at his disposal, but finally in 1911 the law got him. Standard Oil was forced to divest itself of no less than thirty-three separate oil companies. Some of these thrive to this day, in the form of Mobil, Standard Oil, Exxon and Chevron. Capitalism had survived in the land of the free by controlling itself, by being forced to submit to the rule of government. This was not only good for the market, but good for *all* those who competed in it. Ironically, Rockefeller's income actually increased after Standard Oil was broken up. At the start of the twenty-first century, this parable is being repeated. Multinational companies are now growing to the point where they might soon be capable of challenging major governments. For the moment they prefer more devious methods of getting their own way – with one notable exception. As I write these words on my PC, with the aid of my obligatory Windows software, Microsoft stands in the dock like Standard Oil in 1892. Rockefeller was richer than Bill Gates in real terms, but the government had the will and tenacity to hold on for nineteen years. It now remains for us to see what government in the twenty-first century is made of. Capitalism only survives by adapting to the free market. Paradoxically, the free market only survives because its freedom is under control.

At the turn of the twentieth century, in a land where 'enterprise' meant sticking the knife in first, and 'consolidating the market' meant running your opponents out of town, 'consumption' was perhaps bound to be excessive. The victors in this economic war knew no restraint. Social niceties such as class and taste belonged in the Old World. To be new rich in the New World was to enter an entirely novel, yet curiously primitive society. The tribal behaviour of this native society was to be perceptively analysed by yet another eccentric Scandinavian. Thorstein Veblen was a Norwegian American who firmly believed that the neo-classical economists had been wrong in seeing rational self-interest as the prime motive in economic behaviour. On the contrary, people were driven by far more basic psychological forces such as greed, fear and conformity.

Veblen was brought up in a Norwegian farming community in Minnesota during the 1860s. This remote territory had only become a state in 1858, when new immigrants began occupying the land left by displaced Sioux tribes. Veblen didn't learn to speak English until he went to college, and always retained a distorting Norwegian drawl. He habitually wore hand-me-down clothes, long hair and a beard. Even when he progressed east to take a Ph.D. at Yale, he still insisted upon wearing a coonskin Davy Crockett hat. Yet it was evident to all who encountered him that his oddness went far beyond his appearance. He was solitary by nature, temperamentally alien to all forms of human society. It was impossible to avoid his piercing eyes. What exactly lay behind these, and the sarcastic supercilious manner he adopted, it was impossible to tell. After completing his thesis on Kant, Veblen returned to Minnesota: no one was interested in employing this brilliant but difficult oddball. At home he simply retired to his room and read *for seven years*. All attempts to interest him in useful activity on the farm were ignored. It appeared that, having studied philosophy, he had decided to become a philosopher. But his reading of philosophy soon turned his mind to more social questions. The oddities of social behaviour soon became a topic of intriguing interest to the detached eye of the odd observer.

Veblen decided that economics was the most influential aspect of any society. Yet he quickly realized that the economics which in reality moulded society bore little relation to the prevailing theoretical model proposed by neo-classical economists. Mathematical descriptions or models based on rational behaviour ignored the vast range of historical, psychological and anthropological forces at work. In 1888 Veblen briefly broke off his studies to get married to a young woman called Ellen Rolfe, who belonged to the richest family in the region. Veblen may have been a strange-looking loner, but there was something about him that was highly attractive to certain women. Despite his preference for his own company, he was not above turning this attraction to practical use when the mood suited him. Veblen's tolerant but increasingly exasperated father hoped that this marriage would lead to a well-paid job in the Santa Fe Railroad Company, which was run by Ellen's uncle. Yet Veblen showed little interest in the railway business, and continued with his day-long reading.

In 1892 Veblen managed to obtain a low-paid teaching post at the University of Chicago, which had been founded that year by Rockefeller. He was to remain here for fourteen years, his difficult personality and increasing penchant for philandering ensuring that he was never promoted above the level of 'instructor'. His lectures consisted of a bewildering cornucopia of erudition, peppered with quotations ranging from current Chicago slang to medieval Latin hymns – all delivered in his customary sarcastic, deadpan drawl. According to a later colleague, he was 'a satanically detached person who was engaged in laying bare, with utterly cold technique, the diseased nervous sytem of a patient for whom he had no personal feeling beyond his interest in the ailment'. The patient in this case was nothing less than society itself, and the book he wrote on this subject was to make his name.

Veblen's *The Theory of the Leisure Class* was published in 1899. This renegade work took the insights of Darwin's theory of evolution, as well as recent American advances in psychology and anthropology, and applied these to economic life – especially as exhibited by wealthy American society. The irony is evident right from his opening words: 'The institution of a leisure class is found in its best development at the higher stages of the barbarian culture.' The rich believed that they were different from others. Their wealth was no accident of existence; instead it was a reflection of their biological superiority. Their inventiveness, their intellectual gifts, their sheer brilliance and *savoir faire* – these were what set them apart from others. According to Veblen, this delusion had always been the case. (And indeed this quaint myth of social Darwinism lingers on to this day in such out-of-the-way spots as Monaco, Palm Beach and St Moritz.) Veblen proceeded to compare late-nineteenth-century New York high society to the Stone Age existence of the Polynesian islanders and Viking tribal behaviour in the Dark Ages. Tribal leaders in all three of these primitive societies were free from menial activities. While other, lower forms of life worked, they lived effortlessly, doing nothing. Leisure was their defining quality. But in late-nineteenth-century New York people were often too busy working to notice the rich going about their leisure. Superiority has to be demonstrated. The rich found 'the only practical means of impressing one's pecuniary ability

*Thorstein Veblen, the maverick thinker who studied the
primitive tribal behaviour of the rich*

... is an unremitting demonstration of ability to pay'. Caviar,
racehorses, jewels – especially jewels. Tribal leaders in primitive
societies had always adorned their women with expensive trinkets.
The leisured rich possessed, and felt the need to demonstrate that
they possessed, two priceless entities. These entered the language
in Veblen's timeless phrases: 'conspicuous leisure' and 'conspicuous consumption'. As Veblen put it, 'in the one case it is a waste
of time and effort, in the other it is a waste of goods'. The
whole idea of spending money was to impress others. Making the

255

neighbours jealous only increased one's sense of self-importance. The leisure class didn't ostentatiously spend money on useful things – that was for those who needed money, not those who were above it.

The primitive leader demonstrated his superiority with ostentatious examples of his social prowess – giving extravagant feasts, possessing lots of over-adorned wives, featuring in self-glorifying ceremonies. As Veblen observed, 'modern survivals of prowess' remained in the form of ever-changing fashions, cliquey aesthetic tastes, expensive sports such as polo and ocean yacht racing, and ephemeral grand balls set amidst gold-painted cardboard 'Roman' temples or ice sculptures of the host's locomotives. On occasion the cigars on offer would simply be wrapped in $100 bills: guests with panache used these to light their cigars, others retained the wrapping in order to 'inhale wealth'.

The Theory of the Leisure Class was a huge success with all but the class who had most time to read it. The leisure class was immune to envy – such emotions belonged to those unfortunates less gifted by evolution than themselves. However, the mockery of the vulgar masses was another matter altogether. The rich were not used to being a laughing stock. Veblen had written the socio-economic equivalent of Hans Christian Andersen's 'The Emperor's New Clothes'. The rich were found to be wearing nothing but their own delusions of grandeur.

Veblen's insights made a crucial contribution to society's economic self-understanding. Beneath his veneer of eviscerating wit and social observation, Veblen was making some profound criticisms of neo-classical economics. Consumers did not simply purchase goods because these gave utility – fulfilled need, gave satisfaction, pleasure and so forth. They often purchased goods for purely sociological reasons, such as keeping up with the neighbours. 'With the exception of self-preservation, the propensity for emulation is probably the strongest and most alert and persistent of economic motives proper.' And far from giving pleasure, this could even plunge the consumer into debt. We all know someone who just *had* to buy certain clothes, even if they couldn't afford them, because they wouldn't be seen dead in anything less. Fashion, peer pressure, self-esteem and social practice – all these factors, and more, play a part in consumption. This insight of Veblen's would launch a thousand advertising campaigns. What he discovered, we regard as obvious: a sure demonstration of lasting truth.

All this blew a hole in the central neo-classical assumption that consumer spending was a matter of rational calculation. Will product A give me more utility than product B? Neo-classical models – from Walras' equilibrium to Marshall's supply and demand diagrams – all depended upon the consumer asking himself this question, *and answering it rationally*. Such models also depended upon seeing the consumer as an individual, his choice and his calcualtion being *his own*. This was evidently not the case if the consumer's choice was influenced by the choice of others – by fashion, the wish to keep up with others, the need to demonstrate superiority, and, to an increasing extent, by the embryo industry of advertising. Neo-classical computations were based upon false premises: the individual who calculated his own utility, figures derived from an aggregate of such individuals making rational choices. Economics sought to understand human behaviour, but Veblen showed that such behaviour could not be understood simply in terms of sophisticated economic laws.

Veblen's second major work, *The Theory of Business Enterprise*, was published in 1904. This struck even more deeply at the heart of accepted economic theory. Neo-classical economics was too formal, too static. It missed the point that economic systems evolve qualitatively as well as quantitively. They transform themselves, becoming structurally different; they don't just expand, producing ever-greater quantities. The US Steel Corporation owned by Carnegie was not simply a vastly expanded company factory such as Walras had used as his model. It was a vast conglomerate consisting of many differing units of production. Veblen also noticed that within such newly evolved economic structures contradictory forces had begun to emerge. As the manufacturing side of the business became potentially more efficient due to new industrial processes, so the leadership of these vast enterprises became increasingly irrational. The engineers were the ones responsible for the new technology, and left to their own devices they would have produced sufficient goods for all. But they were frustrated in this aim by the businessmen who ran these organizations, who were fearful of disrupting the markets and undermining prices. In Veblen's words, 'Safe and sane business management . . . reduces itself in the main to a sagacious use of sabotage.' The robber barons were more interested in fighting each other than in producing more goods. They sought to gain power by ruining their opponents, rather than running their businesses more efficiently. Quite simply, near-monopolies such as US Steel and Standard Oil were outstripping all moral control. Was what succeeded in business right or wrong? Veblen astutely noted, 'It is not easy in any given case – indeed, it is at times impossible until the courts have spoken – to say whether it is an instance of praiseworthy salesmanship or a penitentiary offense.' The reality was that the ways of society had little or nothing to do with morality or truth. They simply reflected the prevailing way of thought amongst the powerful. Two centuries previously Mandeville had demonstrated that economic virtues were the opposite of Christian virtues. Later, more liberal times had tended to concur with Dr Johnson's remark: 'There are few ways in which a man can be more innocently employed than in getting money.' Now even blinkered literary types could no longer believe in this illusion. The truth of Mandeville's *Fable of the Bees* had become more blatant than ever before.

Neo-classical thinkers saw economics as the analysis of optimum distribution under equilibrium. What was necessary to bring about a state where all markets cleared? This was essentially concerned with stasis, but according to Veblen real economics should be concerned with *destabilizing* influences. These came about through technological change, and also through changes of taste. As Veblen himself foresaw, entire industries would be made or broken by a shift in fashion. Years later, it took nylons replacing silk stockings to popularize synthetic fabrics in clothing. Each new president of the US is approached by the clothing industry, which tries to induce him to wear a hat – a move which would immediately return this item to male fashionable wear.

Veblen also instigated a variant of institutional economics. Previously Marx had seen economic change as resulting from the conflict between social institutions such as classes, between corporations and trade unions, between government and those governed. Veblen concentrated on the less visible institutions which determined how economics itself worked, in other words, 'the settled habits of thought common to the generality of men'. Economic institutions consisted of 'usage, customs and canons of conduct'.

In the end Veblen's unorthodox behaviour became too much for the University of Chicago, and in 1906 he was asked to leave. His works had made him famous, but their gleeful undermining of the neo-classical attempt to transform economics into a serious science had made him few friends in academic circles. From now on he was forced to take on a series of lowly posts in remote universities. Here, he persisted in his unusual personal behaviour and wayward studies, producing works such as *The Theory of Women's Dress* and *The Intellectual Pre-eminence of the Jews in Europe*. He understood the outsider's view of society, and the vital role such elements played in economic life. At the University of Missouri he lived alone in a bare cellar for a year, entering and leaving by a skylight to avoid human contact. Yet he continued to attract the intense interest of a succession of female students and faculty wives, one of whom described him as a 'chimpanzee'. Whether this referred to his hirsute appearance or his sexual behaviour, and whether it was intended as a compliment or a slur, remains unclear.

In 1926, at the age of sixty-nine, Veblen gave up teaching and retired to a remote wooden hut 2,000 feet above the sea on the coast of California. Here he died on 3 August 1929, less than three months before the Wall Street Crash, which would transform economic thinking for ever. During the ensuing Depression, Veblen's sardonic anti-establishment stance attracted an increasing number of readers. People preferred descriptions of the iniquities of economic reality to theoretical models of a system which had broken down.

But how could this come about? How could capitalism simply break down? The omens of this coming event were ambiguous. The early economic thinkers had noticed the occurrence of 'partial gluts' in the market, when there were too many goods and not enough purchasers. But according to Say's Law, 'supply creates its own demand'. In the end the market would always right itself: free trade was essentially self-adjusting. Say's early-nineteenth-century contemporary Malthus had disagreed. These 'partial gluts' would continue to recur, until in the end there would be a 'general glut' which would bring the entire economy to a standstill. Economists had warily continued to observe the cycles which business underwent. Jevons's study of business cycles and sunspots had been eccentric only in its conclusion. His inquiry into the economics of business cycles, and what precisely they involved, pointed the way ahead.

The first comprehensive analysis of business cycles was carried out by Schumpeter, the renowned American economist of Austrian origins. Joseph Schumpeter was born in 1883 in the old Austro-Hungarian empire. His father owned a textile factory, and young Joseph was sent to the Theresianum, the elite school in Vienna, then on to the University of Vienna, which at the time contained the leading school of economics in Europe. Right from the start Schumpeter made no secret of his ambitions. He wished to become 'a great lover, a great horseman and a great economist'. He set about this task with aristocratic panache – a pose adopted to disguise middle-class uncertainties. His first job on leaving university was in Cairo, managing the estates of an Egyptian princess. In between times he dashed off a treatise on the theory of economic development. In both the practical and the theoretical fields he proved a huge success. The princess found her income from her estates doubled; Schumpeter's treatise was greeted as a masterpiece and secured him a professorship in economics back in Austria-Hungary. Over the next decade or so he acquired a vast theoretical learning, an English wife and a reputation for delivering his lectures in riding habit.

The First World War proved disastrous for Austria-Hungary, Europe's largest empire. Defeated, it splintered into over half a dozen national territories. Remnant Austria found itself beset by spiralling inflation and faced with industrial ruin. There was only one hope. The finest economic theorist in the land, the 36-year-old Schumpeter, was appointed minister of finance. He quickly acquired a castle, a string of horses and almost as many mistresses. (Divorce would follow a year later.) But there was also work to be done. Here was Schumpeter's chance to apply his vast theoretical knowledge to a national economy. Amidst widespread surprise, after putting his ideas into practice for just seven months, Schumpeter resigned. When asked why, he replied, 'I have no wish to remain minister of finance for a country which is about to go bankrupt.' Running a national economy was evidently a different matter from managing a few paddy fields along the Nile. Having demonstrated his ability to bankrupt an entire national economy, Schumpeter was immediately appointed president of a private Viennese bank. A few years later the bank would go the same way as the national economy. Schumpeter, who had surreptitiously borrowed large sums of money to speculate on high-risk shares, found himself plunged into debt. However, such financial ineptitude was of no concern to the University of Bonn, who quickly appointed him professor of economics, thus allowing Schumpeter to escape the country. As Schumpeter would beguilingly observe, 'We all of us like a sparkling error better than a trivial truth.' Here in Bonn, such was the sparkling authority of his economic pronouncements that he was eventually poached by Harvard, who offered him the finest academic post of its kind in the land. Schumpeter would remain professor of economics at Harvard through two further marriages and a stable of thoroughbreds until his death in 1950.

Here indeed was a man who was ideally placed to elucidate the ups and downs of economic life. When Schumpeter wrote about business cycles, you felt that he knew what he was talking about. His analysis was not entirely original in its parts, but overall it proved devastatingly perceptive. Schumpeter identified three different types of economic cycle. The first was the Kitchin Cycle, named after the South African mining businessman who first identified it. This cycle recurred every forty months or so. Its driving force was business attempting to gauge future sales. To begin with, businesses would overstock in order to remain one step ahead of rising sales. As growth slowed down, businesses would then severely reduce production to avoid being left with large quantities of unsold stock. Then, when growth picked up, they would once again increase production and the cycle would recommence.

The second cycle was known as the Juglar Cycle, after the French medical doctor and economic theorist who was the first to demonstrate its existence in convincing scientific fashion. Juglar regarded the circular flow of economic life as being much like the flow of blood through the body. The Juglar Cycle lasted between nine and ten years, and is widely regarded as the most important of the three cycles. Even today, when experts refer to the 'business cycle', this is the one they have in mind. The Juglar Cycle starts with businesses beginning to expand, investing capital in the modernization of factories and equipment. This lasts between four and five years. During the second half of the cycle there is no further expansion or investment, but gradually the plant and equipment become worn out. The cycle recommences four to five years later, when this worn-out machinery begins to be replaced.

The third type is the Kondratieff Cycle. This is named after the Russian economic prodigy who at twenty-five became deputy minister of food in Kerensky's short-lived liberal government, which ruled in 1917 between the fall of the Tzar and the Bolshevik Revolution. Kondratieff later became an economic advisor to the Revolution, but inevitably disappeared early in the Stalinist purges of the 1930s. The Kondratieff Cycle lasts between fifty and sixty years, and is caused by the introduction of new inventions and innovations in commercial technology. During times when growth is sluggish, businesses are not willing to risk introducing untried methods. New inventions and innovative techniques are thus ignored, and tend to accumulate for several decades. However, with the onset of increasing economic growth, businesses become more innovative. New inventions and techniques are applied, and a new Kondratieff Cycle begins. Schumpeter traced the passage of the Kondratieff Cycle through the European economy during the previous two and a half centuries. One new cycle had begun with the early Industrial Revolution at the end of the eighteenth century. New inventions such as the spinning jenny, and the many new uses of steam engine power, were the driving forces. Another Kondratieff Cycle had begun in the mid nineteenth century with the boom in railways. A further cycle had been initiated in the early twentieth century by automobiles, the use of electricity and new chemical techniques.

The following figure incorporates the three cycles, demonstrating how they mutually reinforce and reduce their effects to produce a fourth overall cycle of prosperity, recession, depression and recovery. Schumpeter listed these phases as they applied to the early twentieth century:

Prosperity: 1898–1911
Recession: 1912–24/25
Depression: 1926–38

We can match these dates loosely to the figure, assuming 0 as approximately 1900. Doing this, we can see that the graph cycles bear a close resemblance to the historical reality. The years

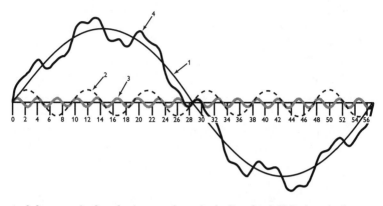

Schumpeter's three business cycles. 1 is the Kondratieff Cycle. 2 is the Juglar Cycle. 3 is the Kitchin Cycle. 4 is the resultant overall cycle produced. Numbers along the horizontal axis are years. Between years 0 and 14 there is rising prosperity. This is followed by recession, which lasts until years 28 and 29. Recession then plunges into depression. This lasts until around year 44, when recovery begins

preceding the First World War were a period of prosperity. Recession hit hard in the early 1920s, and the world slid into the Great Depression during the 1930s. Looking ahead, we can see that the Kondratieff Cycle forecasts the start of a new cycle around 1957. This was matched by the actual prosperity of the early 1950s and 1960s, which saw the wholesale introduction of domestic fridges, washing machines and television sets. Around fifty years later another period of prosperity would be prompted by the computer and telecommunications revolution. However, such booms and economic downturns are very much a matter of economic interpretation. The depression which should have occurred in the 1990s certainly did not match that of the 1930s. Many economists go so far as to deny the very existence of the Kondratieff Cycle. The US economic authority Paul Samuelson even referred to it as 'Pythagorean moonshine'. However, the ten-year Juglar Cycle – simply referred to as the business cycle – remains very much a focus of economic calculation.

Until the early decades of the twentieth century, orthodox economic thought had regarded the role of the entrepreneur as secondary. He merely followed trends, making use of resources to meet consumer demand. Schumpeter chose to see entrepreneurs as the driving force of capitalism. It was they who took the risks which promoted growth. But they also *created* the market. Schumpeter made the amazing claim that most changes in commodities had been 'forced on the consumers'. Initially resistant, consumers had eventually succumbed to 'the elaborate pyrotechnics of advertising'. Examples of Schumpeter's thesis abounded. Initially many people had been reluctant to buy a drink which contained cocaine. But on the crest of a nationwide advertising campaign the sales of Coca-Cola between 1890 and 1900 rose from 9,000 gallons to almost 400,000 gallons. (The cocaine wasn't removed until 1905.) Contemporary advertising has achieved far more spectacular feats. Nowadays we are even willing to pay *more* for the privilege of walking around with *free* advertising logos on our clothes. (In the old days billboard men used to be paid by the hour.)

Schumpeter remained as unpredictable as ever. During the grimmest years of the Great Depression he began writing a major opus predicting the end of capitalism. He decided that – contrary to the evidence visible beyond the palisades of his riding stables – capitalism would not collapse by simply grinding to a halt. Nor would it be torn apart by its own internally generated contradictions, as Marx had predicted, and many believed was then happening. On the contrary, Schumpeter was convinced that capitalism would eventually be destroyed by its own *success*. The present recession was just a 'necessary cold douche', during which outmoded commercial practices would be discarded. According to his business cycle analysis, this was merely the prelude to an upturn which would see a complete restructuring of industry, accompanied by the introduction of a revolutionary range of inno-vations. In the long run, however, capitalism was bound to become the victim of its own success. It would eventually give rise to a highly educated leisure class, who would press for an end to the harsh injustices of the free market and business cycles. They would demand controls, for moral reasons, and the result would be the introduction of socialism. As economist Todd Buchholz has pointed out, for a time this seemed a very real possibility during the social disruptions of the 1960s. Similar forces have begun to emerge in the early twenty-first century – as the disturbances at the World Trade Organization meeting in Seattle and continuing 'anti-capitalist' demonstrations attest. Yet it seems most unlikely that these will result in socialism. Once again, capitalism will have to reform itself, restructuring its moral practices just as it restructures its commercial practices. Here would seem to be another business cycle – this time involving cycles of convergence and divergence. Morality and commercial practice seldom con-verge for long, yet their divergence is never permanent.

Despite the wayward nature of its practitioners, economic theory was now moving ever closer to economic practice.

I I

Cometh the Hour, Cometh the Man

Schumpeter's overall business cycle appeared to accommodate the major historical events of the first half of the twentieth century. The First World War, the Russian Revolution, the Wall Street Crash and the Great Depression, the Second World War – each more or less fitted into place on the cycle. Did this mean that these events were all fundamentally economic in cause? Could they all have been predicted *in economic terms*? Historians continue to argue either way – though it is difficult to believe that all such events can be explained in purely economic terms. History ignores economics at its peril, but the reverse is even more true.

Yet one of these events – the one which would bring the entire world trade to its knees – *was* a purely economic event. The causes of the Wall Street Crash in 1929, and the Great Depression which came in its aftermath, are explicable entirely in economic terms. With the magisterial finality of hindsight, all the experts now agree that this was a disaster waiting to happen. For those on the ground – including the experts – it was another matter. They just didn't see the train until it hit them.

As no less a figure than J. K. Galbraith has put it, 'No one, wise or unwise, knew or knows now when depressions are due or overdue.' As with so many economic cautionary tales, this applies as much today as it did in 1929. During the summer of that year the economic indicators looked good. Business was thriving, and industry still had excess capacity. High profits were coupled with low costs and low wages. The newly elected president Herbert Hoover had promised, 'Poverty will be banished from the nation.'

This growing confidence had encouraged a boom on the New York Stock Exchange. Prices had now been rising almost continuously for over a decade, giving rise to the so-called Great Bull Market. The nation was investing its savings, and more, in shares. Credit was readily available, and cheap. This was the way to make money and everybody was at it – from high-flying financiers to taxi-drivers, from hotel owners to hotel doorkeepers. During the weekends of the long, hot summer of 1929, the Gatsbys of Long Island held open house: the Jazz era was in full swing.

However, a number of perceptive investors had noted that the US economy had reached its cyclical peak. Production, as well as prices, were now heading into a gradual cycle of decline. The acceleration in share prices slowed slightly, but continued to boom. As Galbraith put it, 'Prices were going up because private investors or institutions and their advisors were persuaded that they were going up more, and this persuasion then produced the increase.' Share prices were not about to dip. Even the revered economic guru of the period, Irving Fisher, declared, 'stock prices have reached what looks like a permanent high plateau'. Large and small investors continued to stampede the market, buying all they could lay their hands on. Shares were bought on deposit, and these shares were used as collateral for the purchase of further shares on deposit. There was a proliferation of holding companies (involving credit), shares dealt on the margin (further credit), simple loans – an ever-greater pyramid of credit was being assembled.

Share prices finally peaked in September, followed by a slight dip. The process which had prompted prices to continue rising now began to work the other way. Lack of confidence persuaded further lack of confidence. The modest decline began to accelerate. Then panic set in: suddenly everyone wanted to unload. On Thursday 24 October 13 million shares were traded on the market. The bankers of New York, faced with the same situation as John Law had faced two centuries earlier in Paris, responded in identical fashion. Reassuring statements were issued – the very appearance of which only served to fuel further fears. Word began to spread that the bankers were only issuing such statements to steady the market so they could unload their own holdings. After a weekend of bushfire rumour the Great Bull Market finally changed into the Great Bear Market. On Tuesday 29 October a record 16 million shares were traded: the blackest day in the Exchange's history. Some 300 million shares had been held on the margin: debts were suddenly called in on a colossal scale. Private investors were wiped out by the thousands, their savings transformed overnight into a bottomless pit of debt. Financiers were ruined, directors of long-established stockbroking firms committed suicide. A new image began appearing in the nation's cartoons: the business tycoon preparing to jump from the skyscraper ledge outside his

This Chrysler Roadster cost over $1,500, but the need for ready cash became desperate after the Wall Street Crash plunged thousands of investors into debt

office. As the accountants began going through the books, major embezzlements were uncovered: 'borrowings' which had formed the basis of vast speculations and consequent fortunes. Respected bankers were led off to Sing Sing.

The year 1929 was bad; but the market eventually began to bottom out. The worst was over. Then the market collapsed again. By 1932 New York Stock Exchange prices had sunk to 20 per cent of their former level: $40 billion had simply disappeared into thin air. But this paper reality was stripped away to reveal a bedrock of hard actuality. Nearly 90,000 firms had gone out of business. Of the 25,000 banks in the US, 11,000 had gone bust – taking with them countless small private savings accounts. Industrial output had collapsed by 50 per cent, incomes had also halved, and one-third of the workforce was unemployed. Such figures speak for themselves, though they do not give full voice to the anguish and despair of the millions of individual lives which were ruined. The grim, grey faces of dejection lining sidewalks throughout the land ('Brother, can you spare a dime?'), the haggard humiliation of the soup kitchen queues: here was massive failure in the land of hope. The name of the president who had promised to banish poverty became attached to the shanty towns which sprang up on the outskirts of every city. These were known as Hoovervilles. Entire families were dispossessed of their homes, forced to up sticks and take to the road. As depicted in John Steinbeck's *The Grapes of Wrath*, many headed west for the promised land of California, only to experience further disillusion. Nowhere was immune from the Great Depression.

What took place in America was echoed throughout the world. Following the First World War, it had been plain for all to see that America was now the world's leading industrial power. This would prove true for better and for worse. Post-war European recovery had been heavily financed by American investment. America's woes were soon repeated in Europe. In 1931 Credit-Anstalt, Austria's major bank, collapsed. Germany, struggling to recover from the hyper-inflation of the 1920s, plunged into crisis once more with over 25 per cent of the workforce unemployed. Britain, which had formerly assumed control of the world's exchange markets, proved no longer up to the role. America was unwilling to step into the breach, and international exchange rates became increasingly unbalanced. In a desperate attempt to isolate themselves from the financial epidemic sweeping the world, countries tried to place themselves in economic quarantine. Protective tariff barriers began springing up. By 1933 world trade had slumped by *nearly 70 per cent* . Free trade and liberal democracy appeared to have failed. Strong remedies were needed: strong leaders emerged, promising national salvation. Fascism spread from Mussolini's Italy to Hitler's Germany, then later to Greece and Spain. Many others – throughout the world – saw communism as the only answer. In Russia, Stalin extended his power with a series of purges and show trials.

Economists looked on aghast. What on earth had gone wrong with the market? The Great Depression went on and on – four years, five years, six years – with no sign of recovery. But this just couldn't happen. According to Say's Law, 'supply creates its own demand'. What had become of the invisible hand? According to Adam Smith, and the neo-classical tradition, the market should eventually have righted itself. And what about the business cycle? Did all this mean that Marx had been right after all? Would capitalism finally be destroyed by its own internally generated contradictions? In London Keynes declared,

> We are today in the middle of the greatest economic catastrophe
> – the greatest catastrophe due almost entirely to economic
> causes – of the modern world . . . the view is held in Moscow
> that this is the last, culminating crisis of capitalism and that our
> existing order of society will not survive it.

How could capitalism possibly be rescued? Of all those who wrung their hands over this question, it would be Keynes who came up with the answer to it. In doing so, he would play a major role – both theoretical and practical – in saving the world as we know it. Here would be an economist whose only peers were Adam Smith and Karl Marx.

John Maynard Keynes (pronounced 'Canes') was born in Cambridge in 1883. His family tree could be traced as far back as William de Cahagnes, who had arrived in England in 1066 with William the Conqueror. Like both Mill and Walras, Maynard Keynes's father too was an economic thinker of some originality, lecturing at Cambridge. His mother was a forceful character who would eventually become Cambridge's first woman mayor. Young Maynard was brought up in a high-minded and highly conventional Victorian atmosphere, though he appeared not to find this too stifling. His precocity enabled him to think for himself, but he kept this to himself. Such independence of mind would become a lifelong trait. At thirteen he was sent to Eton, the most elitist school in a class-ridden country. He was a strikingly ugly adolescent, with a damp protruding lower lip and receding chin; his eyes were said to be 'soft as bees' bottoms in blue flowers', but capable of an icy stare of disapproval. His tall, gangling physique enabled him to become a promising rower. He also showed exceptional ability at both maths and classics. Keynes became a popular figure, largely on account of his natural charm and quick wit. He cultivated a wispy moustache and won a scholarship to Cambridge. Here, he studied maths and moral science (philosophy). This combination of ethical inquiry and mathematical science encouraged him to attend some economics lectures by Marshall, who was by now the Grand Old Man of economics. Keynes was enthralled, and was soon pondering the problems of supply and demand equilibrium.

Despite his legendary capacity for work, which would continue throughout his life, Keynes also found time to enjoy himself. (At the end of his life, when asked what he most regretted, he would reply that he wished he had drunk more champagne.) During his years at Cambridge, Keynes made friends with Lytton Strachey, E. M. Forster and Leonard Woolf, future husband of Virginia. These would form the nucleus of the Bloomsbury Group, an intimate circle of artistic and intellectual friends who devoted themselves to the pursuit of truth, excellence and each other. The Bloomsbury Group would have a formative influence on Keynes during the next two decades of his life. In many ways, the group represented a healthy reaction against Victorian stuffiness and conformity, but from the perspective of another century it appears as a peculiarly English anomaly. The 'Bloomsberries', as they were nicknamed, were sexually liberated yet addicted to prurient gossip, free-thinkers yet cliquish in their intellectual fashions, liberal yet elitist, socially and intellectually avant-garde yet incorrigibly snobbish. Lytton Strachey and Keynes found they had both mordant wit and homosexuality in common, and would later begin preying on the younger good-looking members of the group. When it was time to leave Cambridge, Keynes airily declared, 'I want to manage a railway or organize a Trust.' Alas, there were no posts on offer for 23-year-old railway magnates or financial tycoons with no previous experience, so instead Keynes took the highly competitive entrance exam for the civil service. Out of the entire country, he came second – largely owing to his disappointing mark in the economics paper. 'I evidently knew more about economics than my examiners,' he wrote to Lytton Strachey, an arrogant assessment with more than a grain of truth.

Keynes took up a post at the India Office in Whitehall and was bored to death. His first task, which involved months of paperwork, consisted in organizing the purchase of ten pedigree bulls and their shipment to Bombay. The disappointed young captain of industry consoled himself by taking long weekends with his Bloomsbury friends in the country and picking up rough trade on the night streets of London. These disparate social engagements were dutifully listed in his diary: 'To Burley Hill to visit Lytton.' 'Lift boy of Vauxhall.' The latter behaviour exhibited, among other things, a degree of social daring. Homosexuality was still very much illegal and it was only a decade since Oscar Wilde had been publicly disgraced and sentenced to two years' hard labour. In between times Keynes wrote a masterly treatise on probability which criticized – and improved upon – work from De Moivre to Gauss and Edgeworth.

Keynes didn't mince his words. Probability theory as developed by these predecessors was highly mathematical and as such bore little relation to any reality, such as the economy. Mathematical probability pretends to objectivity, it pretends that it knows. 'This event is 90 per cent certain to happen.' Why? Because it has happened 90,000 times out of 100,000. Keynes dismissed 'the so-called Law of Great Numbers', just as he dismissed the use of the word 'event'. The future remains uncertain. There is no *event*. In reality, probability merely forms the starting point in our judgement of likelihood, in our degree of belief concerning the future. We then use logic, reason and even intuition to argue our case. And this arguing is largely done with ourselves – making our belief in what will happen heavily subjective. 'The basis of our degrees of belief is part of our human outfit.' It depends upon our judgement. Even if the probability of something happening is 99 per cent, there always remains the element of uncertainty. The 100–1 outsider in the Derby always has a few backers. And why not? Just a couple of years before Keynes started writing his treatise on probability, Signorinetta had won the Derby at 100–1.

Uncertainty was the most important factor, that which differentiated propositions about the future from other propositions. This uncertainty would later play a fundamental role in Keynes's economic theory. Every aspect of economics was characterized by uncertainty – from the capitalist risking his capital on a new venture, to the way the poorest family decides to spend its income. How many people who buy lottery tickets can't 'afford' to buy one? It is wrong to assess human behaviour as rational. We are driven by 'deeper and blinder passions'. Economic behaviour is human, and thus irrational.

An early draft of Keynes's *A Treatise on Probability* was sufficient to get him invited back to Cambridge in 1908 to teach economics, an unusual appointment considering he had no degree in this subject. The great Marshall was convinced that Keynes would one day take on his mantle. Ironically, Keynes was now turning away from Marshall's ideas on equilibrium, developing new notions which were at variance with the whole tradition of neo-classical economics.

At the outbreak of the First World War in 1914, Keynes's Bloomsbury principles led him to register as a conscientious objector, a principled and socially brave act. Paradoxically, this would result in him playing a greater role in Britain's war effort than many generals. In 1915 he was called to the Treasury. Here, he was entrusted with the task of preserving Britain's vital supply of foreign currency, a feat he performed with considerable success. Inevitably, he agonized over the ambiguity of his position: 'I work for a government I despise for ends I think criminal.' His successes at the Treasury were on occasion achieved with great ingenuity. When France was unable to repay its debts, he suggested that these could be settled in the form of art works. Art prices had plummeted during the war, and Keynes was able to use his Bloomsbury-honed artistic acumen to secure for Britain a number of cut-price modern treasures. These included masterworks by Delacroix, Gauguin and Monet – which remain to this day jewels of Britain's national collection. For good measure, Keynes even acquired a bargain Cézanne for himself.

At the end of the war Keynes was selected as a Treasury representative on the British delegation to the Versailles Peace Conference, which opened outside Paris in 1919. One of the purposes of the conference was to settle the amount of compensation which should be paid to the Allies by the defeated Germans. Keynes watched, with increasing disbelief, the behaviour of the negotiating leaders. The British prime minister, Lloyd George, privately exercised his wily Welsh charm on the other delegates, meanwhile pitching his public pronouncements for maximum effect on the British electorate back home. Keynes summed up his character: 'When he is alone in his room there is no one there.' Keynes's impression of the the US president, Wilson, was little better: 'Like Odysseus, the President looked wiser when he was seated.' The French leader Clemenceau was even worse: 'He had one illusion – France; and one disillusion – mankind.' Clemenceau was determined to extract the maximum punitive reparations from Germany. On the advice of Britain's bankers, Lloyd George connivingly suggested a massive payment of £24 billion, with a high initial payment and the rest spread over a few years. Keynes looked on aghast. But he was only a minor delegate, with no influence whatsoever on the proceedings. He knew that realistically Germany could probably only pay £2 billion, with even this spread over several years. As Keynes put it, 'If Germany is to be "milked", she must not, first of all, be ruined.' He foresaw dire consequences, predicting the 'devastation of Europe'. Germany's initial payments would only serve to raise prices in the receiving countries, reducing their exports. Meanwhile Germany would collapse under the burden of payments, bringing recession and panic.

At Versailles Keynes became ill with worry and strain, owing to his inability to avert the catastrophe of which everyone else seemed blithely unaware. He eventually resigned in disgust, returned to London, and set down his anger in a hastily written, but meticulously argued, book, *The Economic Consequences of Peace*. This was rushed into publication by Christmas 1919. It caused a sensation and made Keynes a household name overnight. Imaginative apocalyptic predictions frequently strike a popular chord, but Keynes was intent on more than doom-mongering. He predicted a depression throughout Europe, leading in some countries – especially Germany – to starvation. But he warned: 'Men will not always die quietly.' Such developments could lead to 'hysteria'.

Instead he proposed war payments which Germany could afford to pay, and the formation of a European free trade area to ensure Continent-wide economic recovery. Keynes would, of course, be vindicated. A succession of German economic disasters followed – from the hyper-inflation of the 1920s (when workers had to be paid with wheelbarrow-loads of paper notes), through to the second economic collapse of the early 1930s which brought Hitler to power. The millstone of German reparations may not have been entirely responsible for these events, but it significantly deepened their effects. (President Hindenberg invited Hitler to take power after he had received only 37 per cent of the votes. Just a few per cent less, and this would have been out of the question.) After the publication of *The Economic Consequences of Peace*, Keynes came to be seen as the leading economic oracle in the land. He was regularly asked to contribute articles to all the leading newsapers – except that bastion of the establishment, *The Times*. Whitehall – the civil service and the government – would never forgive Keynes for his 'treachery'.

During the 1920s Keynes continued putting forward his increasingly original economic ideas, on matters ranging from the real causes of the growing unemployment in Britain to monetary reform. These were listened to, but politely ignored, in London. As Keynes wrote to one of his Bloomsbury pals, 'To debate monetary reform with a City editor is like debating Darwinism with a bishop 60 years ago.' At Cambridge his ideas fared little better amongst the academics, who still clung to neo-classical orthodoxy. Keynes seemed unable to produce an overall theoretical framework for his ideas, let alone a logical refutation of Marshall's time-honoured notions.

Belying his prodigious industry, Keynes always maintained a studied relaxed pose. Here, he is describing the civilized citizen of the modern world, but he is in fact describing himself:

*A cartoon of John Maynard Keynes at the height
of his fame and influence, by the political
cartoonist John Low*

The inhabitant of London could order by telephone, sipping
his morning tea in bed, the various products of the whole earth
and reasonably expect their early delivery on his doorstep; he
could at the same moment and by the same means adventure
his wealth on the natural resources and new enterprises of any
quarter of the world, and share, without exertion or even trouble,
in their prospective fruits or advantages.

The real picture was not quite so relaxed and rosy. When the great economic advisor inherited a few thousand pounds (at the time £2 was a working wage), he decided to follow his own suggestion. Whereupon he sought to 'adventure his wealth on the natural resources and new enterprises' provided by the stock market and currency exchange dealings, and lost the lot. Fortunately, a sympathetic banker baled him out, and he soon got the hang of things – making £500,000 over the next few years. A man in Keynes's position, even lying in bed sipping his morning tea, had his inside sources concerning the 'prospective fruits or advantages' of the market. This was quite legal in those days. Indeed, it formed the backbone of the Exchange: this was how stockbrokers made their money. Needless to say, such practices are unheard of nowadays. At any rate, stockbrokers do get out of bed.

In the evenings Keynes might accompany some Bloomsbury pals to the theatre to see the latest Shaw, or the ballet at the Alhambra. Diaghilev's Russian Ballet were all the rage. After the show, Diaghilev himself, Picasso, Stravinsky and the leading dancers would on occasion be entertained at one of the Bloomsbury Group houses in Gordon Square. It was at such a party that Keynes met the Russian ballerina Lydia Lopokova. To the surprise of his Cambridge colleagues, and the horror of his Bloomsbury intimates, Keynes fell in love. He was by now in his forties, she was in her twenties. Lydia had received the formal education of a professional ballerina (i.e. practically none), but was said to have a 'natural intelligence and a free spirit'. In fact, she couldn't even speak English properly. Her conversation was peppered with hilarious mistakes. 'I dislike being in the country in August because my legs get so bitten by barristers.' These were so frequent that Keynes referred to them as 'Lydiaspeak.' As one of the Bloomsberries observed, 'At first I put her mental age at eight.' But there was no denying that Keynes was besotted with her, and she appeared equally enamoured of him. In Cambridge circles they wondered what on earth had attracted him to a 'show girl'; others wondered what had attracted him to a girl. In 1925 Maynard and Lydia were married. Bloomsbury remained cool; but Keynes's world, as well as his interests, now extended far beyond their cliquishness. His homosexuality appears to have been a 'phase', albeit one that lasted over twenty years. Maynard and Lydia were to remain close for the rest of his life, though there were no children. Whenever they were separated he would write her frequent loving letters – explaining simply, but without talking down to her, what he was doing. Her replies showed a real grasp of his aims, and were unfailingly tender: 'with caresses large as sea I stretch out to you'. She nursed his poor health, was unstintingly loyal, provided him with the domestic calm he appears to have needed and filled in for him on the social round when he was busy. Over dinner she would delight him with the details of her day: 'I had tea with Lady Grey. She has an ovary which she likes to show everyone.'

Four years later the New York stock market crashed, and once again the wise man with his finger on the pulse of the world economy found himself in a financial hole. (One of the very few who got out in time was Joseph Kennedy, who was thus able to invest his millions in more dependable projects such as bootlegging, and make the fortune which would one day assist JFK to win the presidency. A history of luck, crookery and ambition: such are the ingredients of legend.) Keynes was hit hard by the Crash. He was even forced to consider selling his favourite Matisse, and a few other Impressionist works he had on the walls at home. The world plunged into the Great Depression. As the bad times grew worse, no one knew what to do. The economic experts clung obstinately to their old laisser-faire ways. This was all part of the business cycle: things were bound to take an upturn soon. The unemployed would eventually come to realize that they could only find employment if they were prepared to work for less. And businessmen would also come to their senses by cutting their prices to increase sales. It was up to the businessmen and the workers. Any interference with the normal course of events would only make matters worse: it was not for the economists to step in. As Keynes complained, 'I do not know what makes a man more conservative – to know nothing but the present, or nothing but the past.' Yet still the free marketeers stuck to their guns. Things may have been bad at present, but in the long run they would right themselves. Keynes pointed out dismissively: '*In the long run* we are all dead.'

Keynes sat down to write the work which would show the world the way out of its current catastrophe. This would not only provide theoretical justification for its provisions, it would launch an entirely new way of thinking about economics. Keynes would develop an approach to macroeconomics which found its own moral course between the promiscuity of Adam Smith and the puritanism of Marx.

The General Theory of Employment, Interest and Money was eventually published in 1935. According to one of its greatest admirers, the celebrated twentieth-century American economist Paul Samuelson, 'It is a badly written book, poorly organized ... It is arrogant, bad-tempered, polemical and not overly generous in its acknowledgements. It abounds in mare's nests and confusions ... In short, it is a work of genius.' Fortunately its central message was clear enough. What was the situation? In the words of the haunting contemporary hit by Yip Harburg:

> Brother, can you spare a dime? ...
> Once I built a building, now it's done.
> Once I built a railroad, made it run.

How American to *sing* of such a plight: but the song said it all. There were no more railways to build. There was no investment to build more skyscrapers. (The Empire State Building was completed in 1931; indicatively, it would remain the tallest building in the world until 1954.) What was to be done about the vast grey army of the unemployed – with their hands out begging for dimes, pennies, centimes, piastres, annas ... ? Keynes made an only partly satirical suggestion: 'If the Treasury were to fill old bottles with banknotes, bury them at suitable depths in disused coalmines which are then filled up to the surface with town rubbish, and leave it to private enterprise . . . there would be no more unemployment.' There would be work, money to spend, the economy would restart.

His revolutionary solution to the Depression? *Spend* the way out of it. Create work. And with no private investment, this could only be done by the government. Forget about balancing the budget: use money to create public work projects. Instead of burying money in bottles, build roads, schools, hospitals. The newly elected American president, F. D. Roosevelt, the greatest president that country has ever had, put such policy into practice in the New Deal. He was immediately criticized for initiating projects as crazy as bottles down mineshafts. His WPA programme even gave out-of-work writers jobs recording the history of regions throughout the land. This 'useless' project would rescue an integral part of America's self-knowledge. Roosevelt's New Deal had started in 1933. Keynes had long been advocating such measures in newspaper articles, lectures and essays in learned economic journals. His new book would now provide theoretical justification for such revolutionary ideas. He would not only show how they worked, but why they worked. And he would also show why there was no other way which would work. He would argue the world into accepting what came to be known as Keynesian economics.

As he put it in the opening chapter of *The General Theory*, 'The characteristics . . . assumed by the classical theory happen not to be those of the economic society in which we actually live, with the result that its teaching is misleading and disastrous if we attempt to apply it to the facts of experience.' Central to this neo-classical theory, as it is now known, was Say's Law, whereby supply creates its own demand and a full-scale depression was deemed simply impossible. Keynes pointed out that the reverse of Say's Law was in fact the case. It was demand which created supply. As soon as demand dwindled, manufacturers were forced to cut back production and lay off workers. This had a spreading effect. Even one worker laid off meant one entire family cut back on its expenditure.

The shops it used suffered, perhaps even causing another worker to be laid off. As the effect spread through the community, the initial cause acted as a 'multiplier'. On the other hand, when investment took place the multiplier worked in the opposite way. A new machine, a new worker taken on to operate it, more spending in the local shops, a new shop assistant taken on, etc.

According to Keynes the crucial factor in the economy was 'aggregate demand'. As we have seen, this is dependent upon consumer spending and investment. Consumer spending was dependent upon psychology. Anxiety about continuing employment, the state of the country or even a general malaise about The Future all acted as a damper on consumption. People then tended to put away money 'for a rainy day'. This had dire results. When the total demand for goods and services fell below the total income, a recession resulted. Not everything was being returned to the immediate economy and being recycled through it. But surely savings were no longer kept under the bed: they were put into savings accounts which banks could use as loans to finance investment, or they were used to buy shares? The money still circulated. Certainly, but in the short run such savings represented a decrease in demand for consumer goods. This meant unbought products and laid-off workers, and thus these savings became a multiplier. Things didn't look good for the future, business confidence waned. If people wished to save more than business felt confident in investing, this could only result in recession and unemployment.

Consumer confidence, like humanity itself, was not a matter of mathematical calculation. It was not rational. More recent examples of this phenomenon abound. In Argentina in 1978 General Videla's vicious dictatorship was facing economic disaster. Then Argentina won the football World Cup, and national confidence skyrocketed. The consequent consumer binge saved the economy, as well as the loathed Videla. But it can work both ways. In 1999 the Japanese economy showed signs of beginning to emerge from its slump. In an effort to kick-start the economy, the government promised £100 tokens to anyone who spent a similar amount on household or electrical goods. These tokens couldn't be saved: they too had to be spent within a limited period. This looked an even better wheeze than Keynes's bottles-down-the-mine suggestion. Yet there were comparatively few takers, and the economy remained in the doldrums of low aggregate demand. Japanese confidence in the national economy had been shattered by the slump. This confidence would take a lot longer to recover than the economy itself: people continued to save.

Objective measures to improve consumer confidence are usually more orthodox. Lowering interest rates is the favourite ploy. This makes credit more easily obtainable, so that consumers feel more inclined to spend on larger items such as cars or houses. Tax cuts can have a similar effect. But Keynes made it clear that these were not the most important factors where consumer confidence was concerned. Uppermost in people's minds was always income. If people earned more, and felt sure that they would continue to earn in the future, they were likely to spend more and save less. However, Keynes also discovered a 'consumption function', which would affect economic calculations. As people earned more, they spent more: but they tended to spend a lower *portion* of their income. The single young private soldier on weekend leave will spend everything, and more, of his income. The civil servant in the Ministry of Defence will earn more, and spend more, than the private soldier; but he is liable to set aside a portion of his income in savings. The minister of defence himself will earn even more, and spend even more, but a much larger portion of his wages is liable to be converted into savings. Aggregate demand increases as income increases, but it is liable to increase at a decelerating rate.

The other major factor in aggregate demand is investment. This is strongly affected by the interest rates. As Wicksell had shown: if money can be borrowed at 5 per cent, and invested in plant and machinery which gives a yield of 10 per cent, business will be encouraged to expand. Once again, future expectation was important here. Factories do not get built and go into production overnight. If business feels optimistic about the future, it will expand on expectation. Uncertainty leads to a 'wait and see' attitude, which leads to a slowing down of investment, and consequently less aggregate demand. In order to stabilize such swings Keynes suggested a 'socialization of investment'. This has been widely misinterpreted as government control of investment. In practice, it meant government spending to compensate for lack of private investment during times of business pessimism, and a cut-back during periods of optimism so as not to fuel inflation. Amongst other things, Roosevelt's New Deal was also intended to improve confidence, which would then encourage private investment to follow government investment.

In keeping with the 'deeper blinder passions' which guided humanity, investment too was subject more to 'animal spirits' than mere rationality. The multiplier here was like a herd instinct. One businessman decided to invest in new plant. Others noted and followed suit. Confidence bred confidence – as with the self-fulfilling confidence of the New York stockbrokers before the Crash. And of course a similar herd instinct worked in reverse: hence the Crash itself. Keynes's 'socialization of investment' was only intended to rein in this herd instinct, to iron out the peaks and troughs in the business cycle. He had no love for state control communism. On his honeymoon he had travelled to Russia with Lydia to meet her relatives. He had eschewed pessimism, but he had not been impressed with what he had seen: 'Out of the cruelty and stupidity of Old Russia nothing could ever emerge . . . beneath the cruelty and stupidity of the New Russia some spark of the ideal may lie hid.' He believed that 'capitalism, wisely managed [is] more effective than any alternative system yet in sight, but . . . in itself it is in many ways extremely objectionable', a statement which remains as true today as it did then. Yet this for-better-or-worse capitalism depended upon human freedom and energy.

'If Enterprise is afoot, Wealth accumulates, whatever may be happening to Thrift; and if Enterprise is asleep, Wealth decays, whatever Thrift may be doing.' In the end, economy was not the point. 'We threw good housekeeping to the winds. But we saved ourselves and helped save the world.'

Having played his part in saving the world once, Keynes would now be called upon to do it again. Despite having suffered a serious heart attack in 1937, Keynes was summoned back to the Treasury during the Second World War, this time as a senior advisor to the government. The tall, stooping figure with a round face, his moustache now grey above his fleshy lips, became a familiar figure in the corridors of power. He would be observed emerging from private discussions with the chancellor of the exchequer, the governor of the Bank of England, the American ambassador. Since his days at Versailles, he had continued to record his waspish observations of such worthies in his private diaries. Keynes developed the curious habit of assessing people's character according to the shape of their hands. He was particularly fond of his own, and Lydia's (was this perhaps what initially attracted him to her?). Keynes had plaster casts made of his and Lydia's hands: these were kept at home for the delectation of intimate friends. When invited to a meeting with a public figure he had not met before, he confided that 'naturally my concentrated attention was on his hands'. As Keynes sank back in his chair behind his long gangling legs, he would shyly hide his own hands up his sleeves like a Chinese mandarin.

To an increasing extent Keynes became the wartime government's all-purpose economic and financial guru. As ever, he remained an urbane figure. On being appointed a director of the Bank of England (popularly known in the City as 'the old lady of Threadneedle Street'), he commented: 'I am not sure which of us is being made an honest woman.' At the same time he also became a trustee of the National Gallery. In 1942 he was appointed Lord Keynes, an honour which he made fun of to his old Bloomsbury friends, yet secretly it soothed his ego. Lady Lydia announced to her friends, 'Maynard is now a lord and I am a lioness.'

In between his many appointments he wrote on how Britain should pay for its war debt. His analysis of the situation was typically drastic, and typically Keynesian. Once again the laisser-faire approach was lambasted. Without government intervention the decreasing availability of goods would simply lead to inflation. On the other hand, raising taxes would only reduce incomes and cause a recession. Instead, Keynes suggested a scheme for involuntary savings. All income above a fixed level would be held in special bank accounts. The account holder would not be able to withdraw money except in an emergency, but he would receive interest, thus ensuring sufficient circulation of cash to prevent recession. The bank could then loan these savings to the government to pay for the war effort. After the war the money would be made freely available to the account holders, and their consequent spending would play its part in warding off another Great Depression.

This ingenious scheme worked for Britain, but in the long run it did nothing for the new Keynesian economics, which became increasingly tarred with the brush of state control. Critical questions began to be asked. What was Keynes's so-called 'General Theory'? Surely it was nothing more than a series of emergency measures relating to a particular historical period of economic crisis: the Great Depression. It wasn't a *general* theory at all. In the end, neo-classical ideas still held true. Free trade, the fluctuation of supply and demand, unrestricted competition, equilibrium where all markets cleared – these were what mattered. The market without such freedom was nothing.

Yes and no. Keynes insisted that the 'tacit assumptions' of neo-classical economics 'are seldom or never satisfied . . . It cannot solve the problems of the actual world'. Yet according to the final passage in *The General Theory*, 'If our central controls succeed in establishing an aggregate volume of output corresponding to full employment as nearly as is practicable, the classical theory comes into its own again from this point onwards.' He seemed to want to have it both ways. Yet for the time being it was 'central control' that was his aim – and this he set about instituting on a world scale. As the war drew to its close, Keynes criss-crossed the Atlantic, holding meetings in America with the aim of establishing a stable post-war economy, as well as settling the little matter of Britain's colossal wartime debts. No living economist had ever achieved such stature. As the British naval cruiser HMS *Onslaught* passed the liner carrying Keynes to America, its captain signalled, 'Best of luck to you and your distinguished passenger.' Keynes was seen as the world's economic saviour, no less.

In July 1944 delegates from forty-four countries met at Bretton Woods, New Hampshire. Here Keynes masterminded an international plan to avoid another Great Depression. During the 1930s, in an effort to stave off internal depression, individual countries had taken to 'exporting unemployment'. A country would devalue its currency, in order to make its goods cheaper and more competitive on the international market. This would bring employment to its own factories, by transferring unemployment to its competitors. Other countries would then follow suit, in a series of competitive devaluations. This trade war had merely resulted in further depression and mass unemployment – with no one able to buy any goods, no matter how cheap they were. To forestall any repeat of such a disaster, a series of measures was agreed. All countries promised to peg their currency to the price of gold. This meant that all currencies had a fixed exchange rate, one to the other. (This was subtly different from the old gold standard, which had died in the 1930s, whereby a country retained sufficient gold stocks to back its paper currency. The emphasis now was on the exchange rate, rather than the actuality, of gold.)

Measures were also put in place to avoid imbalances in international trade. When a country imports more goods than it exports, it has a trade deficit, which must be paid for – resulting in an outflow of currency. Under the new measures money would be loaned to countries that had a trade deficit, whilst countries that had a trade surplus would be penalized. This latter move would encourage the stronger economic countries to import more goods, thus avoiding another world depression. But the strongest economy of all, the United States, could not agree. Its manufacturing base had not been destroyed by bombing, and it knew it would inevitably run a heavy surplus until world trade got back on its feet again. It saw no reason why it should be penalized for this.

Despite this setback, the Bretton Woods conference agreed to the establishment of a World Bank and an International Monetary Fund (IMF). These would oversee the international flow of money, and make loans to countries in difficulties. Money would be made available to cover trade deficits, but only if appropriate economic policies were pursued to remedy the situation. From now on world trade would remain under a measure of control, in order to forestall any future global economic catastrophe. The establishment of the World Trade Organization (WTO) in 1995 represents just the latest stage in this on-going process of international economic co-operation. This oversees worldwide agreements on tariffs and trade. So why is the WTO attracting such bad publicity at present? What is all the fuss about every time they hold a conference? As it was in the beginning, so it remains. The countries with the most economic clout continue to have the last say in these world organizations. Policies which protect the interests of the bigger, stronger economies as well as the smaller, weaker ones inevitably work in favour of the big boys in the end, even if only in maintaining the status quo. Many favour a system more geared to limiting the power of the strong, and more actively helping the weak.

At the end of the Bretton Woods conference in 1944, the assembled delegates rose to their feet and serenaded Keynes with a rousing chorus of 'For he's a jolly good fellow!' Keynes's pallid, illness-ravaged features are said to have turned pink with sentiment, but he knew all too well that the conference had not gone his way. Keynesian economics may have been imposed upon the world, but this was being done according to the American agenda. Likewise when he negotiated repayment of the huge debt which Britain had incurred to America during the war. The US negotiators insisted upon full repayment. Though the terms were extended and comparatively generous, the fact is Britain was as good as broke. However, America still regarded Britain as a strong competitor in world trade, and wasn't going to give away any advantage. (Astonishingly, it is now known that Keynes's opposite number in these negotiations, senior US Treasury official Harry Dexter White, was in fact a Soviet agent – a weak Britain in Europe was also in the Russian interest.)

When Keynes returned to Britain after renegotiating the British debt, he was criticized on all sides, especially by *The Economist*. According to its argument, but for Britain holding out on its own after the fall of Europe, the Nazis would have won the war. It was Britain which had fought hardest and longest. In moral terms Britain was owed a large debt, not the other way around. Such an argument, though it proved unacceptable at the time, raises a profound issue. This covers not only the way we live, but the economic nature of the world in which we live. On the one hand we are constantly bewailing the fact that almost everything nowadays is seen in terms of money. Everything has its price. On the other hand, we complain that a vast amount of economically beneficial human activity is simply not paid for at all. Bringing up children, maintaining households – much of what used to be called 'women's work' – is simply not classified in economic terms. As a result, women especially tend to be economically under-regarded. Here surely is a case where everything *should* have its price. Britain's 'honour' in the war had no price; yet a person's honour can be priced very highly in the libel courts. The world viewed through the hard realism of economic spectacles still has its blind spots.

The Bretton Woods agreement, with its Keynesian provisos,

would remain in force for the next quarter of a century. During the period from 1945 to 1970 the world would attain lower unemployment rates than at any time during the century. At the same time, the world economy would grow at a rate never seen before in history, achieving an even greater transformation than during the eighty years of the Industrial Revolution (1760–1840). The Bretton Woods era effectively came to an end when President Nixon took the US dollar out of the gold exchange in 1971. (Ironically, it was Nixon who declared, 'We are all Keynesians now'.) Nixon's action meant that the dollar exchange rate floated against other currencies, which soon followed suit in cutting loose from gold. There was now no absolute yardstick: the period of flexible exchange rates had begun.

To a certain extent the dollar has taken over from gold as a yardstick. (The dollar remains the only major currency which has never been devalued in the modern era.) But the dollar is only *a* yardstick, not *the* yardstick. The pound and the yen may rise a similar amount against the dollar, yet the pound may at the same time fall against the yen. Amidst such a complexity of fluctuations the likes of George Soros, the legendary Hungarian currency trader, would make his fortune. Yet at the same time currencies were free to 'find' their own rates of exchange. How? By the same process whereby all other goods found their price on the open market: by supply and demand. There is a 'run' on the dollar (i.e. people will sell their dollars for other currencies) when it is said to be 'overpriced' (i.e. its exchange rates are too high). This leads to an inevitable 'correction' in the price of dollars against other currencies. Such fluctuations can prove a real danger to the world economy when they become violent. No one quite knew what was going to happen when the currencies of the 'tiger economies' in the Far East went into free fall in 1997 – yet the world economy survived. Neo-Keynesian economics would seek to control this, whilst letting minor fluctuations continue in line with neo-classical supply and demand. To repeat Keynes's very words: 'If our central controls succeed . . . the classical theory comes into its own from this point onwards.'

So, contrary to appearances, Keynes did not want to have it both ways. The control he sought was in the macroeconomic sphere. Down at the microeconomic level of individual firms and businesses, he believed the old neo-classical ways should be left intact: market forces should prevail. This leads to the problem of where macroeconomics ends and microeconomics begins. Is it most effective to apply macroeconomic controls at state level, or over larger trading areas such as the European Union? And how much control should be allowed to organizations at the international level, such as the IMF and the WTO? Closer to home: is the government responsible for controlling the national economy, and if so how much should it seek to impose its control? Here Keynesian economics leads us into a region where politics and economics become one. Adam Smith may remain revered but widely unread; Marx may remain reviled but widely unread; Keynes is not only read, but remains a topic of constant debate.

Keynes eventually became exhausted by his constant voyaging across the Atlantic, and his seemingly ceaseless round of complex and vital negotiations. But still he went on. He strove to save the world, to save his country – but not himself. In the course of these negotiations he suffered a number of minor heart attacks. By 1946 the debt question was settled, and Bretton Woods was being put into practice. 'The world is saved,' he declared. He retired to the small estate he had bought in the country for a period of rest. Within a matter of months, he was dead. He was just sixty-two years old. A grateful nation honoured him with a full-scale memorial service at Westminster Abbey, attended by members of the government and representatives of many nations, both large and small. The most moving moment of this service was when his 93-year-old father and his mother were observed making their way slowly up the main aisle towards their seats.

12

The Game to End All Games

Even as Keynes was crossing the Atlantic for Bretton Woods to save the world, others in America were convinced they knew better. Down in Princeton, Johnny von Neumann and Oskar Morgenstern had just put the finishing touches to their 'ultimate solution' to economics: game theory. Morgenstern had already made known his view that Keynes was nothing more than 'a scientific charlatan'. Von Neumann had little time for airing such opinions. In between shuttling from Washington to Princeton, he was also involved in the frantic final stages of assembling the first atomic bomb at Los Alamos. What Keynes was up to was of little concern: von Neumann was too busy contemplating how the Third World War would be conducted to worry himself about the economic consequences of the Second one. His global war strategy, like his solution to economics, would also feature game theory.

As we have seen, John von Neumann came up with game theory in 1928, at the age twenty-five. Yet this was far from being his first great achievement. In fact, it was a minor discovery, compared with his two previous shattering feats – which had unfortunately been accompanied by two equally shattering disappointments. The wunderkind of Budapest, who could crack jokes in ancient Greek with his father at the age of six, had quickly graduated into a fully fledged genius. At the age of eighteen he was simultaneously taking degrees at three different universities – Berlin University (chemistry); the ETH in Zürich, Einstein's former university (chemical engineering); and Budapest University (a *doctorate* in mathematics).

By this period mathematics had arrived at one of the biggest questions in its entire history. How could mathematics be shown to be absolutely certain? How could it be demonstrated that mathematics was a complete logical system, starting with a limited number of simple, self-evident axioms, from which the whole of mathematics could be shown to have been constructed? From the age of seventeen von Neumann was already playing a major role in answering this question – producing papers which aimed at 'axiomizing' mathematics, showing what these basic axioms might be. As his biographer Norman Macrae points out, this was much like a young theology student 'stating in a paper that he intended to give a logically irrefutable proof of the nature of the existence of God'. Seven years later von Neumann appeared to have accomplished his aim. The great mathematicians of the age were overjoyed. Here was proof of the absolute certainty of mathematical truth. Von Neumann had succeeded where all others had failed.

But his triumph was to be short-lived. In 1931 the Austrian mathematician Kurt Gödel produced his famous proof that mathematics was incomplete, and can never achieve absolute certainty. Starting from basic axioms, there will always be certain mathematical propositions which cannot be proved to be true or untrue within that system, unless a further axiom is added. But this only gives rise to further propositions which cannot be proved true or untrue within the system. And so on ad infinitum. Mathematics was incomplete, and Gödel had irrefutably proved it to be so.

Thwarted in his quest for mathematical immortality, von Neumann concentrated his attention on quantum physics. He was now living in Berlin. Einstein and Planck were in residence at the Prussian Academy of Sciences. Meanwhile the new young Turks of German physics led by Werner Heisenberg were 150 miles down the line at the University of Göttingen, home to the great succession of German mathematicians from Gauss to Hilbert. Von Neumann was invited to talks with Hilbert, and would frequently catch the train to Göttingen for the weekend. (Legend has it that more great mathematical problems have been solved on the three-hour train journey between Berlin and Göttingen than anywhere else on earth.)

Quantum physics too had reached a major stage in its development. Experimental evidence showed that the electron could behave like a wave or a particle, two apparently unreconcilable entitites. For a particle is an object, but a wave is merely a motion. Heisenberg had sought to overcome this dilemma by proposing that sub-atomic behaviour simply could not be 'visualized'. There could be no model for the behaviour of particles at this level, such as the electron. Instead, the behaviour of the electron could only be registered as a matrix, filled with figures from different experimental observations. The Austrian physicist Schrödinger objected to this cop-out. He showed how, despite the apparent illogicality, electrons could be described quite adequately in terms of wave-like particles. Their behaviour could be modelled mathematically without resorting to Heisenberg's ridiculous 'matrix'. Though new to quantum mechanics, the youthful von Neumann quickly worked out that both these diametrically opposed methods seemed to produce the same results! In which case, he reasoned, they must in the end be saying the same thing. Bringing the full weight of his mathematical abilities to bear on the problem, von Neumann managed to produce a formula which united these two approaches – a breathtaking achievement, which was only matched by the simultaneous achievement of the English mathematician Paul Dirac, who produced his own formula to link these two approaches. The elegance of von Neumann's formula appealed to the Göttingen mathematicians; the functional advantages of Dirac's formula appealed to the physicists, who simply couldn't understand some of von Neumann's more abstruse maths. It was the physicists who were doing the research, and Dirac's formula was the one which went into use. Once again von Neumann had been trumped in a major discovery. One such opportunity in a lifetime is given to the very few; to lose out on two such opportunities would have been enough to embitter any man.

Yet despite the qualities which would one day make him a model for Dr Strangelove, von Neumann remained an outwardly equable character. Even as a young man he possessed an urbane Austro-Hungarian sophistication, dressing in the silk shirts and well-cut suits one might expect of a respectable banker's son. But Johnny von Neumann was no stuffed shirt: he also enjoyed the louche cabarets and wild parties of 1920s Berlin, while his formidable capacity for cocktails became legendary. Yet nothing, not even alchohol, could penetrate his veneer of sophistication. One was always aware that something else was going on inside his head. Occasionally he would slip away from the social proceedings, in which he had never been fully engaged, and would later be found calculating in some quiet spot. Even in his sleep he calculated. Both his wives would recall his unnerving habit of memorizing the elements of a problem before he went to sleep, and waking up in the middle of the night to write out the solution.

During a summer holiday back in Hungary, von Neumann met up with his friend Nicholas Kaldor, another member of the 'miracle generation' of Hungarians. (Thirty years later Kaldor would become a founder of the post-Keynesian school of economics, and also step into his former mentor's shoes as special advisor to Britain's chancellor of the exchequer.) Von Neumann and Kaldor began talking about economics, a subject unknown to von Neumann, who became fascinated. He asked Kaldor if he knew of any work which presented economic theory in fully mathematical terms. Kaldor recommended Walras' *Elements of Pure Economics*. Here, von Neumann came across the concept of general equilibrium, where supply and demand balanced such that all markets cleared. Von Neumann quickly spotted two errors in Walras' mathematical system. Firstly, to achieve equilibrium, prices would sometimes have to be zero, or even minus quantities. The second error was even more damaging. Walrasian economics, and subsequent mathematical analysis of markets, viewed the situation from an entirely mechanical point of view. The economy was a vast, complex engine. Raise one lever, and this had a mechanical effect on the entire system. (Raising interest rates could be likened to putting a foot on the brake, slowing down the engine of the economy.) Von Neumann rightly objected that economics did not deal with a mechanical situation, it dealt with a human one. The interaction was social, not purely physical. A free-trade economy essentially consists of competitors on the open market. When each competitor acts, he remains uncertain what his competitors will decide to do. Yet every competitor knows that the outcome depends upon *all* their decisions and actions. So any individual's decision will inevitably involve his estimation of how his competitors will act. This, for von Neumann, was the key to economics – not any theoretical mathematical equilibrium. And to account for this situation he came up with game theory.

Von Neumann produced a formula for the best strategy to adopt in any competitive situation, from games to economics. This was the minimax theorem. The competitor should analyse each move he could make, and calculate the maximum possible loss the others could inflict upon him if he made this move. His optimum strategy was then to make the move which could involve him in the minimum maximum possible loss. This worst-case scenario would probably not occur; but if it did, he would suffer less. He would live to fight another day. 'Defeat is inevitable if you aim to win rather than avoid losing'. Here spoke the voice of experience: the man who had lost out twice – to Gödel, then to Dirac – in what for him was the greatest game of all. But this time his victory, or avoidance of defeat, was unassailable. Here was the last word in probability theory. From Pacioli to De Moivre, from Gauss to Keynes – they had merely been grasping at straws. Game theory had the answer to them all.

At this point there emerges a curious parallel between von Neumann's activities in economics and particle physics, one which

*Portrait of John von
Neumann, c. 1950*

may well have influenced his overall conception. In a sense, von
Neumann had been trying to resolve the same problem in both
fields. In physics, light could be reduced to a matrix of purely
mathematical readings, and also to a 'picture' which involved
wave-particles of light. Economics too was proving susceptible
to purely mathematical representation, whilst others insisted on
trying to 'picture' it as a given structure. Ultimately, von Neumann's game theory would attempt to unify these approaches –
not only by providing an overall structure of what was happening
(a competitive game involving many players), but also by enabling
the whole process to be reduced to a set of figures (minimax).

In the early 1930s von Neumann began regularly criss-crossing
the Atlantic, receiving large fees for lecturing at Princeton, as well
as his salary from Berlin, and afterwards Hamburg University.
Then in 1933, at the age of twenty-five, he was appointed as one
of the first permanent professors at the Institute for Advanced
Study, along with Einstein. (Gödel would be given a more junior

post a few years later.) Johnny von Neumann quickly established himself at the IAS, adding a dash of European flair to this monastery of theoretical knowledge. His stylish three-piece suits and colourful ties made a sharp contrast to Einstein's baggy old sweaters and sandals. By now Johnny was married to Mariette, a vivacious scion of Budapest high society. The von Neumann household soon became renowned for its wild weekly cocktail parties, hosted by the rotund, jovial Johnny. He was now beginning to put on weight. As his wife would remark, Johnny knew how to calculate everything except calories. Gourmet cuisine with rich sauces, and Viennese-style cream cakes, remained a perennial favourite.

It was here at the IAS in 1939 that von Neumann was approached by the colourful and ambitious 'grandson of the Kaiser', Oskar Morgenstern. Von Neumann agreed to a collaboration. Together, he and Morgenstern would transform the last word on probability theory into the last word on economics. They began work on the paper which would eventually expand to become the 600-page *Theory of Games and Economic Behavior*. In this von Neumann extended game theory well beyond the comparative simplicities of the minimax theorem.

Game theory could involve two types of game: zero-sum games and non-zero-sum games. In zero-sum games, one player's gain was always the other player's loss (when Sampras wins a set, Agassi loses one). But in non-zero-sum games it was possible for two players to win: the so-called win-win situation – which can result from collaboration, for instance (Sampras and Agassi combine to get the umpire replaced). This can of course result in a lose-lose situation (Sampras and Agassi both disqualified by the umpire). Such strategies could be applied to the economic behaviour of companies competing on the open market, or even consumer choice.

Economics was nothing more or less than a large game involving a given number of players. Companies would collaborate to drive other companies out of the market, looking for a win-win situation (a monopolies commission investigation could result in a lose-lose situation). Individual workers would gather together in unions, also seeking a win-win situation for themselves (a strike which bankrupted the company could result in a lose-lose situation). Game theory was nothing less than a *definition* of what went on in a free market economy.

As we have seen, von Neumann had a withering view of previous economic theory: 'simply a million miles away from . . . an advanced science'. He and Morgenstern sought to create 'something in the truly scientific spirit'. This could only be achieved with game theory, which was 'the proper instrument with which to develop a theory of economic behaviour'. Game theory could provide solid proof for all that had previously been mere conjecture. Such economic theory had been a 'hopelessly unscientific discipline'. They pointed out that 'Before they have been given respective proofs, theory simply does not exist as a scientific theory.' Did this mean that all previous economic theory had been wrong? No: it had simply been a haphazard observation of what took place, without any deeper understanding of what was happening. As von Neumann and Morgenstern explained, 'The movements of the planets had been known long before their courses had been calculated and explained by Newton's theory.' Like Newton's gravity, game theory would change everything. The intention was to provide for economics what von Neumann had so nearly succeeded in supplying for mathematics: a set of basic axioms upon which a structure of irrefutable truth could be built. Economics would be absolute and certain. As von Neumann's biographer, Steve J. Heims pointed out,

> if the von Neumann-Morgenstern belief were indeed justified, the elements of competitiveness and aggressiveness in modern society would be contained within the formalism of a strictly axiomatic mathematical theory, and the problem of 'wise' choice, or 'rational' action, would be reduced to a matter of calculation.

No longer would the world be forced to rely upon mere conjectures such as those proffered by Keynes. (Interestingly, Keynes's strictures concerning mathematical calculations of probability still held true for game theory. The maths was only the beginning of a calculation involving 'our human outfit'. Keynes remained right: in the end we don't calculate our strategy, we choose it. Nemesis in the form of Gödel, then Dirac . . . and now Keynes? Von Neumann appears not to have seen it this way.)

Having dealt with economics, von Neumann went into Dr Strangelove mode. He foresaw the day when game theory would be applied to all kinds of situations. These ranged well beyond economics into politics, foreign policy and even nuclear strategy. (The American science writer Robert Wright even argues in his latest book, *Nonzero: The Logic of Human Destiny*, that game theory can be applied to the entire range of evolution, as well as the unfolding events of history.)

After the successful detonation of the first atomic bombs over Hiroshima and Nagasaki, scientists quickly realized that it would be possible to create an even greater bomb: the hydrogen bomb. Von Neumann, who was one of the few to undertand how precisely this could be done, was all in favour. He lobbied enthusiastically in Washington, and was eventually appointed to head the Atomic Energy Commission. This was the height of the Cold War, and the AEC was seen as the spearhead of America's fight against the Russians. The press was briefed, and greeted von Neumann's appointment with enthusiasm. A jovial, sophisticated von Neumann was profiled in the magazines and described as the 'best brain in the world'.

Von Neumann seized his opportunity. Now was his chance to put game theory into practice. No sooner had the first hydrogen bomb been exploded than von Neumann began urging President Eisenhower to use it on the Russians. Game theory dictated that he strike at once. Fellow nuclear scientist and future Nobel prize-winner Hans Bethe found himself marvelling, and wondering whether 'a brain like von Neumann's does not indicate a species superior to that of man'. A more pertinent suggestion would later be supplied by the psychologist Jacques Ellul. When a man allows his patterns of thought and action to be totally dominated by the 'techo-logic' approach, it is liable to eliminate certain vital elements of his consciousness. Briefly, in Ellul's view such a person is likely to have a diminished awareness of what it means to be a human being. (Von Neumann's wives may well have found themselves pondering on this topic while they lay in bed at night beside a brain calculating in its sleep.)

Was Stanley Kubrick's Dr Strangelove such an exaggeration? Dr Strangelove, the evil genius of the War Room, was willing to abandon the lives of millions in pursuance of his dream. Not for nothing was the full title of Kubrick's film *Dr Strangelove: or How I Learned to Love the Bomb*. Von Neumann remained forever in love with the products of his mind. These were his children who would change the world. However there were others who proved capable of thwarting this egomaniacal ambition. To the list of Gödel, Dirac and Keynes was now added Eisenhower. There would be no 'minimax' first strike against the Russians.

As well as working out how to destroy the world, von Neumann also applied his considerable powers to the problem of how to save it. Another subject which preoccupied his ever-fertile imagination was meteorology. He noted that this was in the same hopelessly unscientific state as economics. What was required was the application of scientific rigour, along with a dose of the customary von Neumann imaginative flair. The world's energy was adversely affected by the polar ice-caps. As he put it, 'The persistence of large ice fields is due to the fact that ice both reflects sunlight energy and radiates away terrestrial energy at an even higher rate than terrestrial soil.' The answer was to spray the polar ice-caps with a highly soluble dark dye, such as potassium permanganate. This would coat the snow purple, so that it wouldn't reflect the sunlight. It would also cause the ice to melt, and the entire planet could then benefit from the ensuing climate change. Iceland would experience the climate of Florida. It doesn't seem to have occurred to the best brain in the world that melting the ice-caps would submerge Florida, as well as coastal cities throughout the globe. But this was not all. Reverting to Dr Strangelove mode once more, von Neumann suggested that meteorology could be put to all manner of fiendishly ingenious uses. But first we would have to learn how to control the weather – a simple matter once meteorology had been transformed into a proper science.

This turned out to be not quite as simple as it appeared. As we now know, meteorology is subject to chaos theory: the butterfly wing in Brazil can cause a tornado in Kansas. It was fortunate that von Neumann encountered difficulties with meteorology which he had not anticipated. But for chaos theory we might have ended up with more than theoretical chaos. Von Neumann was hell-bent on turning the weather into yet another 'ultimate weapon'. As he earnestly explained to the chiefs of staff at the Pentagon, making use of the new scientific meteorology, America could plunge the entire length and breadth of Russia into a new ice age. All this would of course require collecting a vast amount of data, and subjecting this to a series of immensely complex calculations. But there was no need to worry. Von Neumann assured the listening generals that he was developing just the thing to take care of this. In the boiler room at the Institute for Advanced Study his version of the MANIAC computer was already well on the way to completion. Indeed, Johnny von Neumann was lending his unprecedented computational expertise to several similar projects, all of which were gradually piecing together the components for what would become the world's first full-scale computers. (The one being assembled at the top-secret RAND Corporation in California was even named JOHNNIAC in his honour.) But, as ever, von Neumann's thinking was several steps ahead of the field. During a series of lectures at Princeton, he outlined the possibilities of these new computers. What he envisaged was more than just some super calculating machine. In future, MANIAC and its like would lead to the development of a 'theoretical brain'. Apart from having a memory, this would also incorporate a special randomizing feature which mimicked the workings of human imagination. The machine would then possess an intelligence of its own. It could be manufactured with a self-replicating program, so that the computer could reproduce itself. But this device would also include a feature that enabled the next generation of computers to incorporate any improvements, and eliminate any defects, which had developed in the previous generation. These would not only be self-replicating intelligent machines, they would also be a rapidly *evolving* form of intelligence. This would, of course, soon outstrip the intelligence and capabilities of the mere humans who had assembled the first generation of these computers. Von Neumann remained uncharacteristically evasive when questioned about what precisely would happen next.

But none of this was to be. The new ice age, induced by either nuclear or meteorological means, was postponed owing to a tragic circumstance unforeseen by the hero of this tragi-farce. Some time during the summer of 1955 von Neumann suffered a minor accident. While walking down an office corridor, he slipped and banged his shoulder against the wall. The pain persisted and he consulted a specialist. He was found to be suffering from bone cancer, which had already reached an advanced stage.

This bolt from the blue was not without its irony. Von Neumann had scornfully rejected the suggestion that nuclear tests might inflict cancer on the witnessing scientists, amongst whom he was a regular. (Several of the leading scientists who attended the early atomic tests would later die prematurely of obscure forms of cancer. These would include Robert Oppenheimer, who was in charge at Los Alamos; Enrico Fermi, who induced the first nuclear chain reaction; and the quantum physicist Richard Feynman, who began his scientific career as the whizz-kid on the project.)

Von Neumann began working harder than ever, driven by the

thought of all that he had not yet succeeded in doing. Twice he had been thwarted in major achievements, intellectual landmarks which he knew would long outlast the twentieth century. Yet still his overweening ambition remained. There was so much he knew he could do. His 'brainchild' computer was only the beginning. Likewise his ideas about nuclear bombs, ballistic missiles, meteorology, game theory as applied to nuclear 'deterrence' . . . These were but a few of the major intellectual advances which he was known to have been working on. Then, of course, there was economics. The application of game theory to economics would put an end to the woolly unscientific aspects of this field. According to von Neumann, there were just two things needed to complete this task: the establishment of a basic set of economic axioms and the invention of a new type of arithmetic especially designed to build upon these axioms. Just one of these tasks would have daunted even the finest mathematicians alive. But von Neumann felt sure that both tasks were within his capabilities. Hadn't he all but succeeded in the far more difficult task of axiomizing mathematics itself? Hadn't he already conceived of an entirely new branch of mathematics in game theory? All he needed was time.

Within three months von Neumann's cancer was declared incurable: death was only a matter of months away. He was soon reduced to a wheelchair. If the physical effect on von Neumann was catastrophic, the mental effect was horrific. The urbane shell of Johnny von Neumann cracked open to reveal an inner angst-ridden wretch of Dr Strangelove proportions. Steve J. Heims speaks of a 'complete psychological breakdown; panic; screams of uncontrollable terror every night'. His mind was beginning to give way – through a combination of accumulated stress, fear, thwarted ambition and cancer. An aghast Hungarian colleague commented: 'I think von Neumann suffered more when his mind would no longer function, than I have ever seen any human being suffer.' As more than one of his colleagues observed, Johnny von Neumann just didn't know how to die. Having been happily agnostic throughout his life, he turned in frantic fear to the Catholic Church. A Benedictine monk was summoned to give him instruction in the Catholic faith. The von Neumanns had moved from Princeton to a house in Washington, in order to avoid the difficulties of travelling to his government committee meetings. Soon he had to move once more: to a private room in the Walter Reed Hospital.

In the midst of all this von Neumann still found time to ponder on his 'ultimate solution' to economics, though no further written work was produced on this topic for the moment. (Morgenstern was forbidden entry to the hospital, Johnny's wife was most insistent upon this point.) However, von Neumann did have time to complete a paper entitled *The Computer and the Brain*. By now the resemblance to Dr Strangelove was increasing daily, as the crippled figure in the wheelchair was transported from the hospital to meetings of various government committees on nuclear deterrence, foreign policy strategy, intercontinental ballistic missiles and atomic energy. The last – the celebrated AEC – even became known as 'the von Neumann committee'. He also had the ear of President Eisenhower himself. But the cancer continued its relentless progress. By the summer of 1956 von Neumann could no longer be moved, even by wheelchair. Yet still his expert advice was prized above all others. So now the committees came to him. A top-ranking Pentagon official graphically recalled:

> On one dramatic occasion near the end, there was a meeting at Walter Reed Hospital where, gathered around his bedside and attentive to his last words of advice and wisdom, were the Secretary of Defense and his Deputies, the Secretaries of the Army, Navy and Air Force, and all the military Chiefs of Staff. The central figure was the young mathematician who but a few years before had come to the United States as an immigrant from Hungary. I have never witnessed a more dramatic scene or a more moving tribute to a great intelligence.

Yet at night, despite heavy doses of morphine, the raving continued. Screams, confused babble, urgent commands or pleadings, often in Hungarian. An Air Force colonel was assigned to guard him, his orderlies were all hand-picked service personnel with top-secret security clearance, in case his night-raving contained 'classified information of use to a foreign power'. The raving worsened towards the end, yet even then there were moments of calm – when on one occasion the childhood prodigy briefly re-emerged. In order to distract him from his fears, the doctors advised reading to the patient as to a child. They started reading through the 300 pages of Goethe's *Faust* in the original German. One afternoon the reader paused as he turned the page; the seemingly lifeless figure in the bed immediately continued reciting the next lines.

John von Neumann died on 8 February 1957 at the age of fifty-three. He was buried at Princeton on a freezing winter's day. Von Neumann's visibly moved Benedictine confessor conducted a brief Catholic service over the open grave, mumbling some misty words to the grim, hard faces of the assembled top brass, Washington committee chairmen and scientists. The head of Los Alamos turned to his neighbour and remarked, 'If Johnny is where he thought he was going, there must be some very interesting conversations going on about now.'

Epilogue: The Game Goes On

Three billion people had unwittingly survived dooms-day, and life continued. The world economy was by now gradually recovering from the effects of the Second World War. This was to a large extent due to the efforts of one man, a general who almost certainly sat on several of the same committees as von Neumann. The contrast between the compassionate general and the vindictive civilian could not be more stark.

George C. Marshall was born in a small town in western Pennsylvania in 1880. He was the son of a prosperous coal merchant, who subsequently went broke during his son's teenage years. George went to military college and rose to become aide to General Pershing, commander of the American forces in France during the First World War. Marshall was soon being described as 'the greatest military genius since Stonewall Jackson'. By the outbreak of the Second World War, he had risen to become chief of staff of the army. Yet he was to suffer a bitter disappointment when he was passed over in favour of Eisenhower to command the Allied invasion of Europe. In January 1947 President Truman appointed Marshall secretary of state, a post seldom regarded as suitable for the military. Within six months Marshall was speaking at Harvard, delivering the most important economic speech the world had heard since Bretton Woods. He proposed a plan 'against hunger, poverty, desperation and chaos' in Europe. Over the next four years $13 billion was poured into Europe as part of an economic self-help programme, which became known as the Marshall Plan. In some extreme cases financial grants were accompanied by material aid. As with all such well-meaning largesse, there were the usual absurdities. Greece, recovering from famine, was shipped large quantities of cooking oil – when practically the only thing the country could produce was olive oil. The motives of the Marshall Plan were, of course, not entirely philanthropic. Bringing prosperity back to Europe would combat the growing influence of communism. Stalin had occupied the whole of Eastern Europe, and an 'Iron Curtain' now stretched from Bulgaria to East Germany. At the same time, in 'free' Western Europe the largest political parties in France, Italy and Greece were the communists.

European prosperity would also hugely extend the market for American goods. There was no point in being the world's largest and strongest economy when there was no one to trade with. Such carping aside, the Marshall Plan was a huge success. During the period when it was in operation, between 1948 and 1951, industrial production in Western Europe grew by a massive 44 per cent. The relief of suffering behind these figures remains immeasurable. At the end of the war every major European city had been in ruins – with the exception of the few which had been spared for 'cultural' reasons, such as Paris and Rome. By 1951 the Festival of Britain was being held in London, to celebrate the country's emergence from the 'austerity years'. The self-help aspect of the Marshall Plan functioned along Keynesian lines. Aid money was spent with the aim of subsidizing a start-up economy, rebuilding factories and in some cases entire industries. Public projects such as railways, hospitals and schools would also decrease unemployment, so there would be wages to spend on the goods emerging from the new factories. Economic revival in Europe was followed by sustained economic growth, and Truman was soon extending the Marshall Plan to other 'free' (i.e. non-communist) countries throughout the globe. Ironically, it had initially been offered to the whole of Europe, but Russia had vetoed its implementation in the countries behind the Iron Curtain.

On his retirement Marshall's lifetime achievements were

recognized by the great and the good, as well as those at the other end of the moral spectrum. In 1951 he was attacked by the notorious Senator Joe McCarthy and his Unamerican Activities Committee as part of a communist witch-hunt. Marshall had apparently shown that he was a communist sympathizer while chief of staff of the army. He had done this by not ordering American troops to break the Yalta Treaty obligations with the Russians and march straight on to Berlin before the Soviet troops could reach the city – a move which would almost certainly have caused the Third World War to begin before the Second one had ended. In a welcome return to reality in 1953, Marshall was awarded the Nobel prize for peace, the first military man ever to achieve this honour. Six years later he became ill and was moved to the Walter Reed Hospital in Washington. Two years after von Neumann, Marshall died in the same hospital.

The 'greatest military genius since Stonewall Jackson' had refrained from starting the Third World War. The genius who had led even geniuses to suspect that he belonged to 'a species superior to that of man' had done his best to try and start it. Von Neumann had been convinced that game theory would have a similar epoch-making effect on economics – which makes the ensuing episode all the more curious. Von Neumann not only rejected, but failed even to understand, the first major development of game theory. And this development was directly related to economics. In October 1949 von Neumann was approached in his office at Princeton by John Nash, an ambitious 21-year-old maths graduate. Nash's mathematical brilliance had already been recognized, but so also had various other qualities. Nash was six feet tall with a broad-shouldered athletic build and 'handsome as a god'. He was also an aggressive boor who couldn't function without constant demonstrations of his own intellectual superiority. His almost pathological lack of social graces made his repressed homosexuality all the more transparent. At one stage he developed a crush on a fellow student, who was also an exceptional mathematician, though not as exceptional as himself. Nash followed the student around, constantly pestering him with jokey, aggressive remarks, harsh practical jokes, challenging him to competitive mental contests. In a very real sense, Nash was only just in control of himself. He simply didn't know how to live as the person he was. His brash over-competitiveness alienated him from the affection of his colleagues, who nonetheless couldn't help but admire his brilliance. They wished to feed off it, and participate in it, even if it meant courting humiliation. According to his biographer, Sylvia Nasar, 'Nobody remembers seeing Nash with a book during his graduate career.' Yet after lectures given by visiting mathematical high-fliers he would ask obliquely probing questions and appear to develop his own original ideas out of their speculative answers.

Despite his apparent lack of interest in books, Nash must have read *Theory of Games and Economic Behavior* within a couple of years of its publication in 1947. Von Neumann's approach to game theory had been essentially co-operative. As we have seen, its non-zero-sum aspects involved collaboration, producing win-win situations and the like. This very much mirrored von Neumann's character. Many of his greatest intellectual involvements had been collaborative. Even as a youngster he was used to talking out his ideas in the cafés of Budapest and then Berlin. As a young mathematical prodigy in Germany he had worked closely with the great Hilbert in Göttingen. The race to build the first atomic bomb at Los Alamos had been very much a team effort. Even his masterwork on game theory and economics had been written in collaboration with Morgenstern. This was how people worked to bring about optimum situations, in economics as well as mathematics. Game theory understood competition but also realized that the market-place was a social institution involving collaboration.

In keeping with his character, Nash came at game theory from precisely the opposite direction. He lived in a world where there was no real communication. In the struggle for supremacy, for optimum gain, each person acted on his own. Collaborative games were a mere convenience. The game situation was essentially *non-collaborative.*

Nash was the first to distinguish between co-operative and non-co-operative games. He understood that games often involved a combination of both. Players will co-operate with each other for their own gain, but are liable to break off this arrangement/ agreement when it is to their advantage to do so. This was how Senator Joe McCarthy had expected Marshall to act at the end of the Second World War: forget the Yalta agreement and take Berlin while you can. Nash's distinction would open up game theory to a much wider range of applications: here was how the *real* world of economics worked. The railroad magnate J. Pierpoint Morgan had collaborated with his rivals just as long as it suited him. Then he'd switched tactics, ruined them, and taken over their railroads at rock-bottom prices.

Nash knew that he was on to something big, but even he felt daunted at the prospect of explaining his idea to von Neumann. Here would be a 21-year-old graduate student explaining to von Neumann how he had got game theory wrong! Nash was shown into von Neumann's office. The great man himself was seated behind a vast, shiny desk, dressed in a tailor-made suit, waistcoat and silk tie. He appeared more like the head of a big corporation than a professor. But Nash was aware that here was more than just a professor. This was the man who sat on influential government committees, who was a top consultant to Los Alamos and the top-secret RAND Corporation, who even advised the president.

Von Neumann politely indicated for Nash to take a seat, and asked him what he wished to say. Nash started in nervously, while von Neumann listened, his head tilted to one side, his fingers drumming the desktop. Nash had only just begun to outline his idea when von Neumann interrupted him. His lightning mind had already leapt ahead to Nash's conclusion. 'That's trivial,' commented von Neumann dismissively. Nash's point was so obvious that it wasn't worth wasting his time listening to it: von

Neumann's manner made this abundantly clear. There was nothing for Nash to do but get up and leave. His attempt to impress von Neumann had resulted in nothing but a humiliating disaster.

But the humiliation had only been personal. The more Nash thought about it, the more convinced he became that his idea was good. And his crucial new concept was far from being trivial, as von Neumann had claimed. This concept was the equilibrium which could manifest itself under non-collaborative game theory situations. A simple example of this is found in certain types of poker, where all players must act simultaneously. This is precisely analogous to the market situation, where every competitor must act at the same time, without being aware of what strategy their competitors are going to adopt. In a variation on the minimax theorem, Nash worked out how each player can have an optimum (minimum maximum loss) response. When all players adopt their best response, the game – or economic market – reaches an optimal situation. Equilibrium is thus achieved. No player could improve upon his or her situation by using another strategy, given the strategies adopted by the other players.

Making full use of his exceptional mathematical abilities, Nash managed to prove that for a wide range of non-collaborative games – such as economic markets – an equilibrium point does exist. In other words, there is a market situation when all competitors can, with their different strategies, be adopting the right strategy for *them*. This was the equilibrium which von Neumann had mistakenly dismissed as trivial.

Despite his disappointment, Nash developed his ideas into a paper, which was published under the comparatively innocuous title of 'Equilibrium Points in N-Person Games'. Even so, it was immediately noticed – in certain quarters. John Nash was appointed as a consultant to the RAND Corporation in California, a post involving a few highly paid months each summer, which fitted in with his university schedule. RAND (an acronym standing for nothing more than 'research and development') was in fact the clandestine government think-tank in Los Angeles where top brains were set to work out the finer points of nuclear strategy, using game theory and the like. Nash found himself peculiarly suited to what Sylvia Nasar has described as RAND's 'compelling mix of detachment, paranoia and megalomania'. In this highly secretive yet informal atmosphere, Nash finally began to grow up. Yet in his case this involved the development of all his qualities: the good, the bad and the ugly. He became increasingly self-confident yet increasingly eccentric: a curious combination of braggadocio and obliqueness. At last able to admit his homosexual feelings to himself, he took to spending hours on his own ogling the body-builders going through their antics down at Muscle Beach. And it was this which eventually led to his downfall. Homosexuality was still very much a prosecutable offence, even in California. In the early hours of the morning, some time in August 1954, Nash was arrested in a public lavatory by a plain-clothes policeman on homosexual entrapment patrol. He was instantly fired from RAND.

Nash returned to the east coast, where he was now working at the Massachusetts Institute of Technology. Apparently chastened by his experience, he immediately fell in love with a young El Salvadorean woman who was working in the music library. Within two years he was married. But this flight into normality was accompanied by a disturbing drift in the opposite direction. There had always been an unsettling element to Nash's eccentricity, but this now became more apparent. It was nothing spectacular, he was just a bit odder. At the same time his mathematical abilities were reaching the height of their powers. The megalomania encouraged by RAND now emerged in his ambitions. John Nash coldly and deliberately set his sights on winning the Fields Medal. This is the Nobel prize of mathematics, which is, if anything, more difficult to win than a Nobel. After burning his way through three extremely tough problems, Nash finally spotted the big one – a highly abstruse and technical branch of mathematics known as continuity theorem. Amidst bouts of increasingly odd behaviour, he focussed his exceptional energies on producing a proof for this theorem. (In technical terms, according to a colleague, this involved 'extending the Holder estimates known for second-order elliptic equations with two variables and irregular coefficients to higher dimensions'.) Following a number of surprisingly inept starts, Nash finally managed to produce – after six months of intense concentration – a spectacular solution. According to an expert mathematician in this field, Nash's work was nothing less than 'a stroke of genius'. (Again, in technical terms, this time in Nasar's words, 'He approached the problem in an ingeniously roundabout manner, first transforming the nonlinear equations into linear equations and then attacking these by nonlinear means.') Nash's Fields Medal was assured.

Then came the worst possible news. A poor, unknown young southern Italian had published a proof for continuity theorem just a few months earlier in the little-read journal of an Italian regional academy of sciences. This was the work of the obsessive mathematician Ennio De Giorgi, who was described by a colleague as a 'bedraggled, skinny little starved-looking guy'. De Giorgi would later be appointed to the top mathematical professorship in Italy, but would remain unchanged. With no thought for anything but mathematics, he lived a life of monkish poverty, even making his home in his office. Towards the end of his life he became a mystic, devoting all his energies to the attempt to discover a mathematical proof for the existence of God. (During his later years the similarly obsessive Gödel would also succumb to this aberration.)

Nash was shattered by the news of De Giorgi's priority. It was soon acknowledged that Nash's proof was more significant, including as it did some highly original new mathematical techniques – but De Giorgi had published first. There would be no Fields Medal. Not long afterwards it became evident that this was more than a professional catastrophe for Nash. One day he walked into the faculty common room at MIT and held up a copy of *The*

New York Times. He announced to his colleagues that the front-page story contained an encrypted message addressed to him from abstract powers in outer space. The announcement caused little stir. John Nash frequently did this sort of thing: it was evidently another of his aggressive practical jokes. If you challenged him, he was liable to produce a code cipher capable of turning the newspaper column into anything, including the first page of the US Constitution.

Some weeks later he burst into a senior colleague's office and proceeded to draw on the blackboard 'a set that resembled a large, wavy baked potato'. He then added two smaller shapes beside it. This too was not unusual. With grim resignation the interrupted colleague waited for Nash to show off his brilliant mathematical explanation concerning the sets he had drawn. 'This,' Nash indicated the potato, 'is the universe.' He paused, holding his colleague's eye. The colleague returned his gaze without a word: he knew better than to interrupt. With inward impatience he awaited Nash's astonishing mathematical explanation of the contents of these two sets which existed beyond the confines of the universe. Nash pointed to them and announced seriously, 'This is heaven. And this is hell.' The colleague understood at once that no mathematical justification could, or would, be offered for these statements. Nash himself was in another universe.

A short time later, Nash was finally offered a top professorship at the University of Chicago. In his replying letter he turned down the offer, explaining that he would instead be taking up the post of 'Emperor of Antarctica'. Nash's final decline was brutal and speedy – involving quantities of alcohol and bouts of uncontrolled aggression. His wife Alicia had no alternative but to have him committed to a secure mental asylum.

Several years later Nash emerged 'cured', a ghost of his former self. He returned to Princeton, and took to hanging about the campus. Mindful of his former brilliance, and assured that he was harmless, the authorities decided to tolerate his presence. During the 1970s and 1980s he became the local eccentric: a pale, haggard figure in the cafeteria smoking a cadged cigarette, a dazed silhouette sitting motionless for hours on end in the library. Occasionally he might creep into an empty lecture hall and scribble incomprehensible formulae or encoded messages on the blackboard. The students gathered that this rather pathetic figure 'had once been an ace mathematician'.

Meanwhile Nash's name was appearing with increasing frequency in articles in the major academic journals. Applications for his concept of equilibrium in non-co-operative games were being found in fields ranging from sociology to evolutionary biology. The young academics who cited his name generally assumed that he was dead.

Then, some time in 1990, Nash began to recover. The phantom of the library emerged to make contact with former colleagues. They discovered to their surprise that he had started doing 'real mathematics' again. Amazed that he was still capable of original work, one of his former colleagues warily began questioning him. How could he have taken all that stuff about messages from outer space seriously? How could he have believed such nonsense? Nash replied that he had experienced no difficulty in accepting such ideas. Why not? Because they had come into his mind in precisely the same way as his utterly original mathematical ideas.

By the early 1990s Nash equilibrium, as it was now known, was universally acknowledged as a major economic concept. Indeed, it was seen as the biggest step in the application of game theory to economics since von Neumann and Morgenstern. This recognition was finally confirmed in 1994 when the 66-year-old Nash was awarded the Nobel prize for economics.

In the words of Princeton economist Avanish Dixit, 'At last we are seeing the realization of the true potential of the revolution launched by von Neumann and Morgenstern.' Despite such claims, game theory hasn't 'solved' economics by turning it into a hard science, even if it has opened up 'terrain for systematic thinking that was previously closed'.

Long after his death Keynes still remained the driving force in economic theory. And even when the inevitable counter-attack was launched, it didn't come from game theory. The era of the great economists – such as Adam Smith, Marx and Keynes – seemed to be over. In the words of the distinguished American economic historian Robert Heilbroner, 'It is great issues rather than great names that characterize our times . . . in the main, the problems of modern capitalism must be studied *as* problems and cannot be wrapped up in the thought of a single personage.' The next problem which capitalism faced would take it into a post-Keynesian era which has lasted to this day. This problem was called inflation.

When prices began to rise after the Second World War, this was seen as a sign of economic recovery. Once the business cycle began its downward curve this trend would soon be checked. But it went on. And when the postwar boom of the 1960s passed into the recession of the 1970s, prices *went on* rising, in some cases even more rapidly than before. Here was an entirely new problem. Suddenly unemployment was leaping up, and some economies were soon actually producing less than at the end of the Second World War. This should have brought prices to a halt – yet still they continued to rise. Inflation spread like wildfire, and was soon affecting free-market economies throughout the globe, no matter what different measures were tried to combat it. Of the major economies, Britain was the worst hit. In the early 1970s the annual inflation rate more than doubled from just over 3 per cent to almost 8 per cent. Four years later this had leapt to an alarming 25 per cent and rising. Even Japan ended up with a figure of over 15 per cent, while America managed with difficulty to peg the new disease to 10 per cent. (Expanding Third World countries such as Brazil and Mexico were soon experiencing such figures on a *monthly* scale. Countries such as Chile clocked up annual figures of over 1,000 per cent. By 1985 Bolivia had reached a mind-boggling *35,000 per cent.*)

Inflation had moved into a self-perpetuating spiral. Price rises brought demands for wage rises, which in turn caused manufacturing costs to rise, thus causing further price rises, followed by further wage demands ... Under such circumstances, unemployment soared. As a result, the economy began to stagnate. And a new word entered the headlines: stagflation. Previously, stagnation would bring inflation to a halt; and inflation would lift the economy out of stagnation. The politicians wrung their hands in despair.

Finally, one man came forward with an answer. This was the American academic economist Milton Friedman. From his base at the University of Chicago, Friedman insisted there was only one cure for inflation: monetarism. This involved controlling the money supply. Nothing else. Inflation only arose when there was too much money in the economy. In a free market, when too much money was chasing too few goods, prices were bound to rise – during periods of recession just as much as during periods of growth. In Friedman's favourite slogan, 'Only money matters.' Any other form of government interference couldn't in the end halt rising unemployment or stop businesses going bust. All such attempts to control the economy only produced a temporary palliative effect. They were in fact counter-productive: they merely staved off the inevitable, and worsened the effect in the long run. Government spending of all kinds should be cut as far as possible. The public sector should be broken up and privatized, so that its separate elements could be run like businesses that had to survive in the market-place.

Here was laisser-faire at its most extreme. The market was god. This decided success or failure for countries, currencies, firms big and small, even individuals. This was the ultimate game, the most fundamental game of them all: the survival of the fittest. Monetarism was to prove a bitter medicine. Its social effects were brutal, especially in terms of unemployment, and it remains for ever associated with the 1980s conservatism of Ronald Reagan and Margaret Thatcher. It worked, but at a price. Some still question whether it was worth it. Others agree with Friedman, who insisted there was no alternative.

Milton Friedman was born in Brooklyn in 1912. His parents were poor Jewish immigrants from an eastern province of the Austro-Hungarian empire, remote in every sense from the culture of compatriots von Neumann and Morgenstern. Soon after Milton's birth the Friedmans moved to New Jersey, where his mother ran a dry goods store beside an overhead railway bridge in a poor neighbourhood. His father appears to have been the classic pill-popping, angina-ridden small business failure. 'Financial crisis was a constant companion,' Friedman remembers. He has a habit of linking his early memories to present economic circumstances. 'Like many immigrants we lived above the store – the same phenomenon that we observe today in the San Francisco Chinatown near our current residence.' Such observations are pertinent: economic history is still being lived around us, and not only in the ever-changing social kaleidescope of the United States. The Friedmans only survived by paying with post-dated cheques. 'In many undeveloped countries, such as Taiwan in 1962 ... post-dated checks [were] a major and extremely convenient and flexible form of credit instrument.' Family life was sparse but close for Milton and his three sisters. Their father died when his son was fifteen. A year later Milton won a part-scholarship to Rutgers College, in those days a small private university. In order to pay his way Milton worked during the mornings in a nearby restaurant, gulping down his free lunch afterwards to get to college in time for lectures. Bolting his food became a revealing habit which would last throughout his life.

Harsh and conservative ideas are only nourished in those who succeed in escaping such grinding circumstances. While Milton was gaining top grades in mathematics and economics, he was also absorbing the realities of business life. A year after he entered college came the Wall Street Crash. The owner of the restaurant where he worked adapted and the business survived. Then he sold it, and business nose-dived under the new owner – who eventually sold it back to the old owner at a loss, whereupon business picked up again. As someone who survived on tips, Friedman remembers this well: 'That cycle was repeated at least once more during my tenure: under the control of the right person, a booming business, under someone else, a dismal flop.' All this only seems to have reinforced the more brutal lesson from his father's experience: 'I have long believed that a feeling for economics is something people are born with rather than acquire by education. Many highly intelligent and even highly trained professional economists know the words but don't get the music.' Some would argue that Friedman ended up dancing along with the music all too readily, without paying sufficient attention to the heartbreak of the words.

Friedman went on to do graduate work at the University of Chicago, and in 1935 ended up in Washington, compiling consumption statistics for the Roosevelt New Deal. This job embodied what would become two important elements in Friedman's approach. His theories would always be backed with strong empirical and statistical evidence. Likewise, when he attacked Keynesian economics, he would know it from the inside: he had actually helped run the New Deal. As he would later admit, 'I came to be among the best-known critics of the growth in centralized government that the New Deal initiated. Yet ironically the New Deal was a life-saver for [me] personally.' Despite his admission, the full extent of the irony does not seem to have penetrated. This anomalous situation would be repeated after the Second World War, when Friedman worked in Paris for the Marshall Plan. Here a curious historical parallel came about. In Paris after the First World War, Keynes had predicted that the terms of the Allied reparations would result in disaster. And they did. In Paris after the Second World War, Friedman predicted that the terms of the postwar Allied agreement would end in disaster. And these did too. At Bretton Woods Keynes had overseen the fixing of international exchange rates. Friedman now asserted that this had been a mistake. Countries would begin to suffer from balance of payments crises, and would increasingly have to be baled out by the International Monetary Fund. (And he was right: by the 1970s even Britain would have to apply for a large loan from the IMF. But by then Friedman had already been vindicated. Nixon had decreed that the dollar exchange rate against gold should be allowed to float, and the era of flexible exchange rates had just begun.)

It wasn't until the 1960s that Friedman began the rise to fame which would make him a household name – first in America, and then throughout the world. In 1966 he started writing his regular column in *Newsweek*, commenting on public policy. This platform for his opinions and ideas would last until the mid 1980s. By then he had also acted as advisor to two presidents (Nixon and Reagan), and been awarded the 1976 Nobel prize for economics. In 1980 he published *Free to Choose*, one of the very few cutting-edge works on economics ever to become a best-seller. In this he argued for the end of practically all government controls. An exaggeration? Friedman advocated the abolition of wage, rent and price controls; the end of minimum-wage legislation, public education and government grants for universities. At the same time, he also advocated cutting governments down to size by selling off – 'privatizing' – antediluvian, unmotivated state monopolies. Nothing was secure – not even social security. This too had to be cut.

Here indeed was cutting-edge economics. For those prepared to listen, his arguments had their own persuasive logic. 'Government measures that promote personal equality or equality of opportunity enhance liberty: government measures to achieve "fair shares for all" reduce liberty.' He then asks: 'If what people get is determined by "fairness", who is to decide what is fair?' Friedman's answer to this question was 'the market'. However, there was another answer, which was perhaps beneath a Nobel prize-winner's notice. In a democracy it is the people who decide what is fair, when they vote in the government. Admittedly, this is a far from perfect system – but it would seem better than most of us surrendering ourselves to something closely resembling fate (manipulated by the self-interested gods of the market-place). Any government which followed Friedman's advice to the letter, and abolished its powers in such wholesale fashion, would find the voters completing this process at the next election. Even Reagan and Thatcher were forced to accept this. The economics practised by politicians is limited by the reality of the ballot box. Economics for Friedman has no such restraints. For him, economics is what 'works'.

This said, Friedman's arguments demanded – and still demand – a hearing. Their economic 'music' is very compelling, as even those who don't like the words are forced to concede. This is the siren song that has changed the world. Friedman begins at the beginning. For him, Adam Smith's key insight was this: exchange between two parties is voluntary and only takes place if both believe they are going to benefit. 'Most economic fallacies derive from the neglect of this simple insight, from the tendency to assume there is a fixed pie, that one party can gain only at the expense of another.' Economics is a win-win situation. We interfere with this at our peril. Friedman quotes Justice Louis Brandeis: 'The greatest dangers to liberty lurk in insidious encroachment by men of zeal, well-meaning but without understanding.' This statement was made in 1928 and would surely have been retracted in the ensuing years of the Great Depression. Not so, says Friedman. Immediately prior to the Wall Street Crash of 1929 the Federal Reserve had reined in the money supply by keeping interest rates high, because they were worried about speculation on the stock market getting out of control. However, in the period following the Crash they left the interest rates high. This kept money supply low, spending plummeted, and the Depression bit. In Friedman's words, 'The Great Contraction [of money supply] is tragic testimony to the power of monetary policy ... not as Keynes believed evidence of impotence.' The situation was only worsened when people needing money sought to withdraw funds from their savings accounts. Insufficient availability of money meant that banks soon began failing at an alarming rate, and the Depression bit harder still.

Expanding the money supply will inflate the economy. (More money: more spending, businesses investing in expansion, etc.) Similarly, restricting the money supply will combat inflation. (Less money: less spending, lower prices, etc.) This is done by monetary policy, where the central bank controls interest rates and the domestic money supply. In a direct contradiction of the Keynesian approach, Friedman advocates monetary policy over the alternative fiscal policy, where the government attempts to control the economy by means of taxation and public spending. Monetary policy has proved highly effective in dealing with inflation. Hence the enormous importance attached to the pronouncements by Alan Greenspan of the Federal Reserve – these concern the lowering or raising of interest rates, making money more or less easily available.

At the same time Friedman recommended that the central bank should increase the money supply (actual amount of money in circulation) by between 3 and 5 per cent each year, thus keeping it in line with the normal growth rate of the US economy. This would enable spending to increase without fuelling inflation. Meanwhile his policy of flexible exchange rates enabled the contagion of inflation to be contained on the international scene. With fixed exchange rates countries simply exported inflation. Prices rose at home, so cheaper foreign goods were purchased. This led to increased spending and a rise in prices in other countries, leading to inflation there too. With flexible exchange rates, when prices rose at home the value of the currency went down. Thus, it wasn't cheaper to buy foreign goods, inflation would not be exported, and local industry would pick up.

But the conquest of inflation had its price – most notably in slow growth and unemployment. Friedman identified a 'natural rate of unemployment', above which inflation would begin to rise. As Friedman himself conceded in his Nobel speech, this natural rate was not 'a numerical constant'. It could rise steeply. Yet all attempts to reduce unemployment below this rate, no matter how well-meaning, were futile. Such action would simply result in temporary relief. Increased employment would bring about higher prices, which would result in lower spending, and then lower production. Inflation would thus return, and bring about an inevitable rise in unemployment back to the natural rate and probably higher. Once again, this was seen as economics at the expense of people. High 'natural' unemployment rates covered an actuality of widespread social deprivation. Figures in statistical documents translated into broken marriages, alcoholism, drugs, despair, sink estates, rising crime levels, underclass alienation and all the rest. People with no hope aren't just going to stand there and be counted.

Low inflation, a slow-growing but perfectly functioning economy and mass unemployment. People voted in governments to *act*. According to Friedman, the market itself performed the only effective action. It was as if nothing had happened since Adam Smith's invisible hand worked its miracle (which even he had had his misgivings about). But Friedman stuck to his guns. His was the classic liberalism which Smith had also espoused, or so he claimed. He was no conservative dinosaur. 'Differences in economic policy among disinterested citizens derive predominantly from different predictions about the economic consequences of taking action . . . rather than from fundamental differences in basic values.' This may have applied in the blessed isles of academia, but on the small island of Manhattan – with its opposing ghettoes in Wall Street and Harlem – it felt very different. Greed and need don't make for shared 'basic values'.

Finally, Friedman has also made an important contribution to the methodology of economics. When constructing an economic model, with the aim of predicting the outcome of a situation, certain assumptions always have to be made. For instance, the risible '*homo economicus*' – a mere conduit pipe of consumption – frequently takes the lead role in the drama of the markets. Likewise, a single firm will be isolated from the storms of the macroeconomy, in order to devise an efficient microeconomy. Such assumptions can be extremely unrealistic. Yet for Friedman this does not matter. As he points out, all assumptions are bound to be unrealistic: models will inevitably be reductionist. What matters is whether the theoretical model makes predictions which prove to be true. Is the theory supported by consequent economic data? If this is the case, then it doesn't matter how unrealistic are its assumptions. Friedman agrees with Keynes's mentor Marshall: economics is 'an engine for discovering the truth, not a branch of mathematics'. Or even an exact science. It is what works.

Friedman's ideas have continued to predominate on the economic scene, always combating any humanistic drift back towards Keynesian compassion. Keynes saw the world as he would like to see it; Friedman would argue that he sees the world as it is. So where do we go from here? At present there are many outstanding economists working in a wide variety of fields. Yet none shows any sign of having a transcendent effect – as Keynes did in the early and mid part of the twentieth century, and Friedman did in the later part. The next Adam Smith, the new Ricardo, another Marx – where will he or she come from, what will he or she do? Is there indeed any place left for such a transcendent figure? Have we finally arrived at that point which Heilbroner identified, where all that is left are *problems* to be solved, rather than a vision to be enacted? Individual visions are dangerous things. Keynes may have had his humane vision of how the world should be, but Marx also had his vision. Yet what could the new Adam Smith do? What is left for him to do? For better or for worse, this will only be discovered by the next Adam Smith – who, we can be sure, will be very different from the previous one. Perhaps at this very moment he is sitting in his wheelchair, pocket calculator in hand as his black-gloved, synthetic finger jabs at the keys. The rim of his shaded glasses glints in the aritificial light as he mutters to himself, his guttural chuckles barely audible above the hum from the bank of computers at his back.

Sources

Because this is intended as a popular work, I have not included footnotes and an exhaustive list of sources. Where appropriate, the sources of most direct quotes have been indicated in the text. The following is a list of suggested further reading, for those who wish to follow up on the main subjects included in each chapter.

Prologue

John von Neumann by Norman Macrea (Pantheon, New York 1992): the latest biography, giving all the jolly and gruesome details.

Chapter 1: Something Out Of Nothing Comes

Exposition of Double-Entry Book-keeping by Luca Pacioli, trans. Antonia von Gebsattel (CLI, Venice, Italy 1994): the relevant chapter from Pacioli's master-work. An underestimated, and now largely unread, tract which arguably had more effect on history than Marx's *Communist Manifesto*.

Short biographical sketches of both Graunt and Petty appear in: *Brief Lives* by John Aubrey (Penguin, London 1989). Although contemporary, these portraits are not always reliable.

Chapter 2: The Richest Man in the World

There are several biographies of John Law. The most recent is
The Moneymaker by Janet Gleeson (Bantam, London 1999). This
is a good popular account of John Law's life and times. Those
who prefer a more scholarly, detailed approach may prefer *John
Law* by Antoin E. Murphy (Oxford University Press, Oxford
1997). This deals in some depth with the economic aspects of
the Law fiasco.

Chapter 3: Before Adam

A History of Economics by J. K. Galbraith (Penguin, London 1990):
the opening five chapters give a good overview of economic
history up to and including the period covered in this chapter.
Sadly there is no biography of Johann Becher in English. Further
details of this remarkable man may be found in *Mendeleyev's
Dream* by Paul Strathern (Hamish Hamilton, London 2000).
This concentrates on his scientific activities.

Chapter 4: The Founding Father

The Life of Adam Smith by Ian Simpson Ross (Oxford University
Press, Oxford 1995): the latest full biography giving a highly
sympathetic portrait with many period details and good exposition of his ideas.

Chapter 5: French Optimists and British Pessimists

Population Malthus by Patricia James (Routledge, London 1979): reliable full-length biography which also includes many details of his friend Ricardo.

There is no full-length biography of Ricardo, but a good biographical sketch can be found in: *Essays in Biography* by John Maynard Keynes (Macmillan, London 1972).

Chapter 6: Brave New Worlds

Henri Saint-Simon: Selected Writings edited by Keith Tylor (Croom Helm, London 1975): a good account of his ideas.

Robert Owen of New Lanark by Margaret Cole (Batchworth Press, London 1953): one of the few reasonably reliable records of his life.

Chapter 7: The Pleasure Principle

Bentham by John Dinwiddy (Oxford University Press, Oxford 1989): a good introductory work, explaining his ideas and their effects.

Autobiography by John Stuart Mill (Penguin, London 1989): a highly revealing, if not always fully accurate, account of his life and ideas.

Chapter 8: Workers of the World Unite

Karl Marx by Francis Wheen (Fourth Estate, London 1999): a superb biography, giving all the salient details of his life, times and ideas.

The Communist Party Manifesto by Karl Marx and Friedrich Engels (Junius, London 1996): a highly readable incitement to revolution, including the best short account of their ideas.

An expanded version of this chapter, with emphasis on Marx's philosophical thought and works, appears in my *Marx in 90 Minutes* (Ivan Dee, Chicago 2001).

Chapter 9: Measure for Measure

Men of Mathematics by E. T. Bell (Penguin, London 1965): contains a highly readable chapter on the life and work of Gauss.

Essays in Biography by John Maynard Keynes (Macmillan, London 1972): contains a good brief account of Edgeworth and his ideas.

Fifty Major Economists by Steven Pressman (Routledge, London 1999): gives the most easily obtainable, but a very brief, account of Wicksell's life and ideas.

Chapter 10: Into the Modern Age

Essays in Biography by John Maynard Keynes (Macmillan, London 1972): contains the best short account of Marshall's life and work, by his most distinguished pupil.

The Theory of the Leisure Class by Thorstein Veblen (Dover, Mineola NY 1994): that rarity, a work on economics that can be read for wicked pleasure.

History of Economic Analysis by Joseph Schumpeter (Allen and Unwin, London 1986): his massive masterpiece which makes an interesting alternative to the usual histories of economics.

Chapter 11: Cometh the Hour, Cometh the Man

The Great Crash of 1929 by J. K. Galbraith (Deutsch, London 1980):
a fascinating in-depth account of what went wrong, and the
spectacular events themselves.

Maynard Keynes: An Economist's Biography by Donald Moggeridge
(Routledge, London 1992): the best single-volume biography
of his life, times and ideas.

Chapter 12: The Game to End All Games

Prisoner's Dilemma by William Poundstone (Oxford University
Press, Oxford 1993): a detailed account of the development
of game theory, interwoven with biographical details of von
Neumann and Morgenstern.

A Beautiful Mind by Sylvia Nasar (Faber, London 1998): a superb
full-length account of the life of John Nash, giving fascinating
details and utterly comprehensible mathematics. The finest
mathematical biography for years.

Epilogue: The Game Goes On

Two Lucky People by Milton and Rose Friedman (Chicago University Press, Chicago 1998): their joint autobiography, which tells
the pleasant side of the story, as well as giving away enough for
you to judge for yourself.

Free to Choose by Milton Friedman (Penguin, London 1980): the
controversial best-seller which outlines his views, no holds
barred.

For a general overview on the present economic situation try:
Butterfly Economics by Paul Ormerod (Faber, London 1998)

Index

Page numbers in italics denote illustration